UNDER THE MICROSCOPE

SCIENCE TOOLS

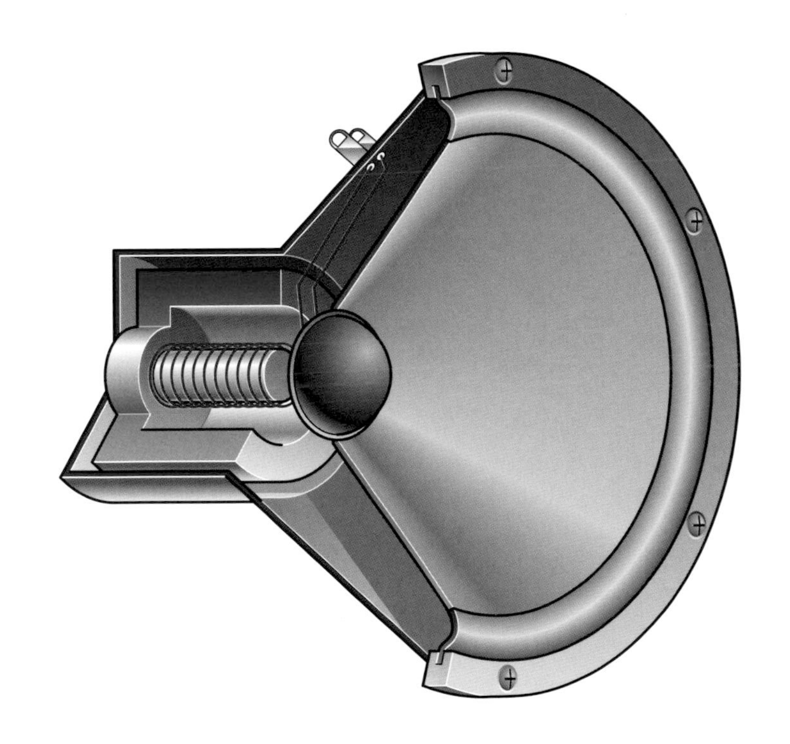

7 USING SOUND

John O.E. Clark

About this set

SCIENCE TOOLS deals with the instruments and methods that scientists use to measure and record their observations. Theoretical scientists apply their minds to explaining a whole range of natural phenomena. Often the only way of testing these theories is through practical scientific experiment and measurement—which are achieved using a wide selection of scientific tools. To explain the principles and practice of scientific measurement, the nine volumes in this set are organized as follows:

Volume 1—Length and Distance; Volume 2—Measuring Time; Volume 3—Force and Pressure; Volume 4—Electrical Measurement; Volume 5—Using Visible Light; Volume 6—Using Invisible Light; Volume 7—Using Sound; Volume 8—Scientific Analysis; Volume 9—Scientific Classification.

The topics within each volume are presented as self-contained sections, so that your knowledge of the subject increases in logical stages. Each section is illustrated with color photographs, and there are diagrams to explain the workings of the science tools being described. Many sections also contain short biographies of the scientists who discovered the principles that the tools employ.

Pages at the end of each book include a glossary that gives the meanings of scientific terms used, a list of other sources of reference (books and websites), and an index to all the volumes in the set. There are cross-references within volumes and from volume to volume at the bottom of the pages to link topics for a fuller understanding.

Published 2003 by Grolier Educational, Danbury, CT 06816

This edition published exclusively for the school and library market

Planned and produced by Andromeda Oxford Limited,
11-13 The Vineyard,
Abingdon, Oxon OX14 3PX

Project Director Graham Bateman
Editors John Woodruff, Shaun Barrington
Editorial assistant Marian Dreier
Picture manager Claire Turner
Production Clive Sparling

Design and origination by Gecko

Printed in Hong Kong

Library of Congress Cataloging-in-Publication Data

Clark, John Owens Edward.
Under the microscope : science tools / John O.E. Clark.
p. cm.
Summary: Describes the fundamental units and measuring devices that scientists use to bring systematic order to the world around them.
Contents: v. 1. Length and distance -- v. 2. Measuring time -- v. 3. Force and pressure -- v. 4. Electrical measurement -- v. 5. Using visible light -- v. 6. Using invisible light -- v. 7. Using sound -- v. 8. Scientific analysis -- v. 9. Scientific classification.
ISBN 0-7172-5628-6 (set : alk. paper) -- ISBN 0-7172-5629-4 (v. 1 : alk. paper) -- ISBN 0-7172-5630-8 (v. 2 : alk. paper) -- ISBN 0-7172-5631-6 (v. 3 : alk. paper) -- ISBN 0-7172-5632-4 (v. 4 : alk. paper) -- ISBN 0-7172-5633-2 (v. 5 : alk. paper) -- ISBN 0-7172-5634-0 (v. 6 : alk. paper) -- ISBN 0-7172-5635-9 (v. 7 : alk. paper) -- ISBN 0-7172-5636-7 (v. 8 : alk. paper) -- ISBN 0-7172-5637-5 (v. 9 : alk. paper)
1. Weights and measures--Juvenile literature. 2. Measuring instruments--Juvenile literature. 3. Scientific apparatus and instruments--Juvenile literature. [1. Weights and measures. 2. Measuring instruments. 3. Scientific apparatus and instruments.] I. Title: Science tools. II. Title.
QC90.6 .C57 2002
530.8--dc21

2002002598

About this volume

Volume 7 of *Science Tools* takes us into the world of sound, which is a type of energy that travels as pressure waves, usually in the air or through water. We can hear a range of sound frequencies using our ears. Microphones imitate this action, converting sound waves into varying electric currents that can be "captured" and stored on magnetic recording tape or on a compact disk. Such recordings can be replayed using an amplifier and loudspeakers. The pitch of ultrasound is too high for human ears, though scientists make use of it in sonar and various scanning devices. Low-pitched infrasound, at the other end of the scale, is too low to hear, but it can be felt in the tremors produced by earthquakes.

Contents

4 Main units of measurement
6 UNITS AND PROPERTIES OF SOUND
10 DETECTING SOUND
16 MODERN SOUND RECORDING
20 REPRODUCING SOUND
24 ULTRASOUND—TOO HIGH TO HEAR
28 UNDERWATER ECHOES
32 INFRASOUND—TOO LOW TO HEAR
40 Glossary
42 Set Index
48 Further reading/websites and picture credits

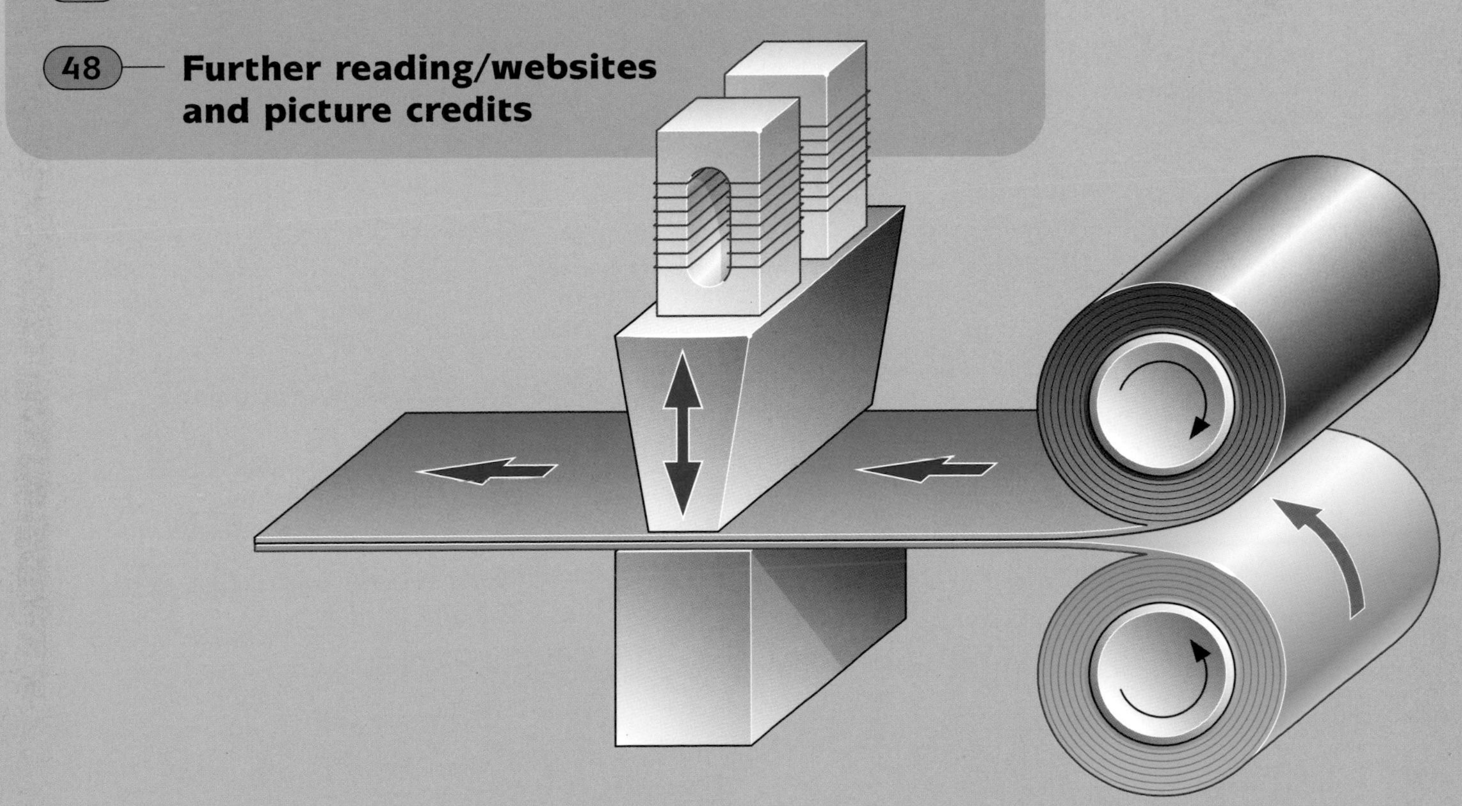

Main units of measurement

Scientists spend much of their time looking at things and making measurements. These observations allow them to develop theories, from which they can sometimes formulate laws. For example, by observing objects as they fell to the ground, the English scientist Isaac Newton developed the law of gravity.

To make measurements, scientists use various kinds of apparatus, which we are calling "science tools." They also need a system of units in which to measure things. Sometimes the units are the same as those we use every day. For instance, they measure time using hours, minutes, and seconds—the same units we use to time a race or bake a cake. More often, though, scientists use special units rather than everyday ones. That is so that all scientists throughout the world can employ exactly the same units. (When they don't, the results can be very costly. Confusion over units once made NASA scientists lose all contact with a space probe to Mars.) A meter is the same length everywhere. But everyday units sometimes vary from country to country. A gallon in the United States, for example, is not the same as the gallon people use in Great Britain (a U.S. gallon is about one-fifth smaller than a UK gallon).

On these two pages, which for convenience are repeated in each volume of *Science Tools*, are set out the main scientific units and some of their everyday equivalents. The first and in some ways most important group are the SI units (SI stands for Système International, or International System). There are seven base units, plus two for measuring angles (Table 1). Then there are 18 other derived SI units that have special names. Table 2 lists the 11 commonest ones, all named after famous scientists. The 18 derived units are defined in terms of the 9 base units. For example, the unit of force (the newton) can be defined in terms of mass and acceleration (which itself is measured in units of distance and time).

▼ **Table 1. Base units** of the SI system

Quantity	Name	Symbol
length	meter	m
mass	kilogram	kg
time	second	s
electric current	ampere	A
temperature	kelvin	K
luminous intensity	candela	cd
amount of substance	mole	mol
plane angle	radian	rad
solid angle	steradian	sr

▼ **Table 2. Derived SI units** with special names

Quantity	Name	Symbol
energy	joule	J
force	newton	N
frequency	hertz	Hz
pressure	pascal	Pa
power	watt	W
electric charge	coulomb	C
potential difference	volt	V
resistance	ohm	Ω
capacitance	farad	F
conductance	siemens	S
inductance	henry	H

▼ **Table 3. Metric prefixes** for multiples and submultiples

PREFIX	SYMBOL	MULTIPLE
deka-	da	ten (×10)
hecto-	h	hundred ($\times10^2$)
kilo-	k	thousand ($\times10^3$)
mega-	M	million ($\times10^6$)
giga-	G	billion ($\times10^9$)
PREFIX	**SYMBOL**	**SUBMULTIPLE**
deci-	d	tenth ($\times10^{-1}$)
centi-	c	hundredth ($\times10^{-2}$)
milli-	m	thousandth ($\times10^{-3}$)
micro-	µ	millionth ($\times10^{-6}$)
nano-	n	billionth ($\times10^{-9}$)

Scientists often want to measure a quantity that is much smaller or much bigger than the appropriate unit. A meter is not much good for expressing the thickness of a human hair or the distance to the Moon. So there are a number of prefixes that can be tacked onto the beginning of the unit's name. The prefix milli-, for example, stands for one-thousandth. Therefore a millimeter is one-thousandth of a meter. Kilo- stands for one thousand times, so a kilometer is 1,000 meters. The commonest prefixes are listed in Table 3.

Table 4 shows you how to convert from everyday units (known as customary units) into metric units, for example from inches to centimeters or miles to kilometers. Sometimes you may want to convert the other way, from metric to customary. To do this, divide by the factor in Table 4 (not multiply). So, to convert from inches to centimeters, *multiply* by 2.54. To convert from centimeters to inches, *divide* by 2.54. More detailed listings of different types of units and their conversions are given on pages 6–7 of each volume. You do not have to remember all the names: They are described or defined as you need to know them throughout *Science Tools*.

▶ **Table 4. Conversion** to metric units

TO CONVERT FROM	TO	MULTIPLY BY
inches (in.)	centimeters (cm)	2.54
feet (ft)	centimeters (cm)	30.5
feet (ft)	meters (m)	0.305
yards (yd)	meters (m)	0.914
miles (mi)	kilometers (km)	1.61
square inches (sq in.)	square centimeters (sq cm)	6.45
square feet (sq ft)	square meters (sq m)	0.0930
square yards (sq yd)	square meters (sq m)	0.836
acres (A)	hectares (ha)	0.405
square miles (sq mi)	hectares (ha)	259
square miles (sq mi)	square kilometers (sq km)	2.59
cubic inches (cu in.)	cubic centimeters (cc)	16.4
cubic feet (cu ft)	cubic meters (cu m)	0.0283
cubic yards (cu yd)	cubic meters (cu m)	0.765
gills (gi)	cubic centimeters (cc)	118
pints (pt)	liters (l)	0.473
quarts (qt)	liters (l)	0.946
gallons (gal)	liters (l)	3.79
drams (dr)	grams (g)	1.77
ounces (oz)	grams (g)	28.3
pounds (lb)	kilograms (kg)	0.454
hundredweights (cwt)	kilograms (kg)	45.4
tons (short)	tonnes (t)	0.907

Units and properties of sound

Sound has two main properties—pitch and loudness. Pitch is how high or low a sound is. It is related to frequency, which is the number of sound waves that pass by in one second. Frequency is measured in hertz. The loudness or volume of a sound is measured in decibels.

Volumes 5 and 6 of *Science Tools* deal with light and the other kinds of radiation that together make up the electromagnetic spectrum. All those kinds of radiation have something in common—they travel as waves, transmitting energy, and can even cross the empty vacuum of space. But there is another source of energy that travels as waves, though it cannot cross space, and it has to have something, a "medium," to travel in. That energy is sound. From the quietest whisper to the roar of a military jet airplane, all sounds reach our ears as waves traveling through the air.

Properties of sound

Like all waves, sound waves have various characteristics—"properties," in the language of scientists—that can be measured. As you can see from the upper diagram on the opposite page, a wave is a series of *crests* and *troughs* traveling in a particular direction. The distance between two neighboring crests—or troughs—is the *wavelength.* Sound waves that we can hear vary in wavelength from several centimeters to a few millimeters.

▼ **A marching band** makes a variety of sounds. The small brass instruments produce the high notes, while the larger ones sound lower notes.

Wave measurements

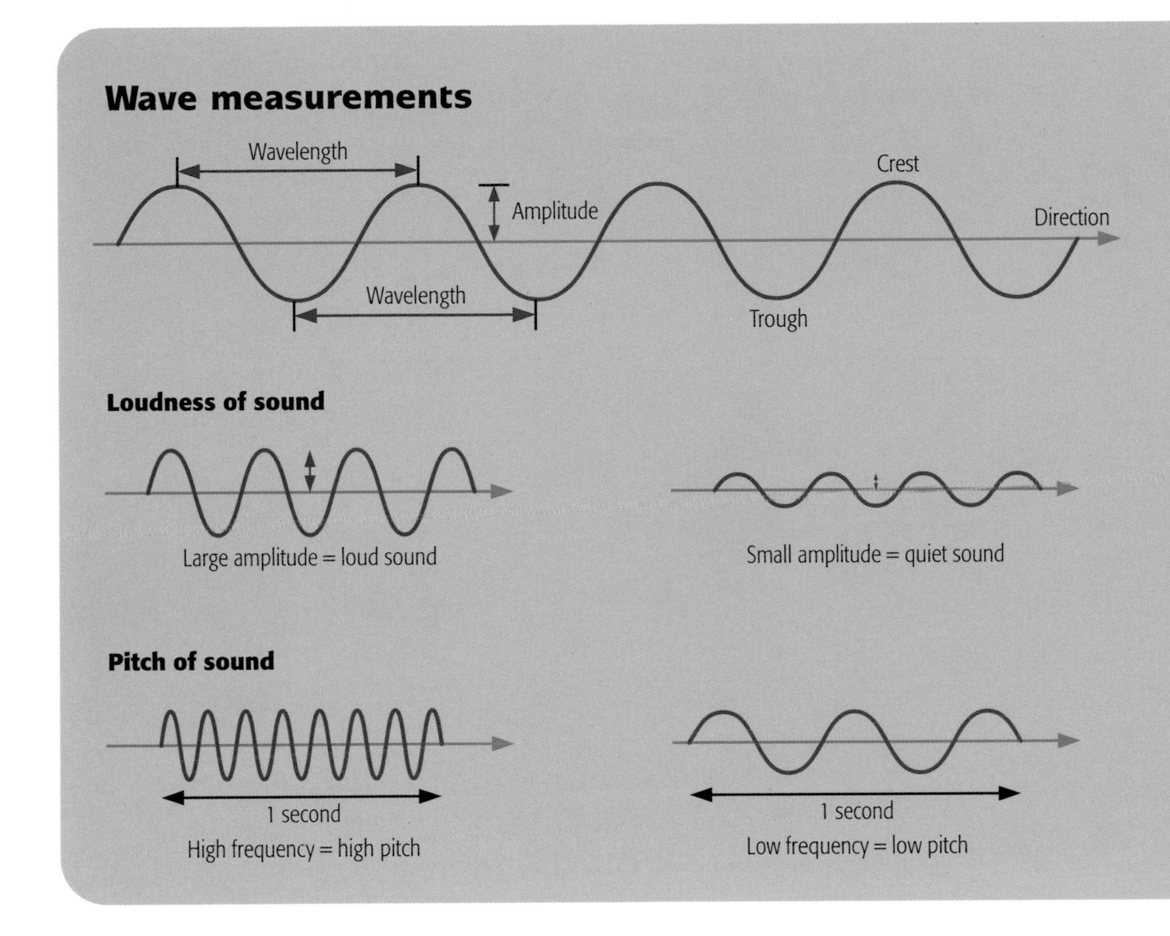

Wavelength is the distance between two successive crests (or two successive troughs) of a sound wave. Amplitude, a measure of loudness, is the maximum height of a wave. Frequency is the number of crests (or troughs) per second.

The maximum distance of a wave from its average position—the height of a crest or the depth of a trough—is its *amplitude*, as shown in the middle diagram. Amplitude is a measure of the loudness or *intensity* of a sound. A large amplitude corresponds to a loud sound, a small amplitude to a quiet sound. Sound intensity is measured in units of watts (energy) per square meter (area), written W/m^2. The quietest sounds we can hear have an intensity of about a million millionth of a watt per square meter ($10^{-12} W/m^2$). An intensity of $1 W/m^2$ is so loud it can damage our hearing.

Because of the way the human ear works, doubling the intensity of a sound does not make a sound that is twice as loud. So scientists use another unit, called the decibel (symbol dB), for measuring sound loudness. The decibel scale is based on logarithms, so it does not increase in equal steps in the way that the graduations on a ruler do. The quietest sound we can hear—called the threshold of hearing—is given a decibel value of 0. A 10-decibel sound is ten times as loud, a 20-decibel sound is a hundred times as loud, and a 30-decibel sound is a thousand times as loud as a 0-decibel sound. The noise of a children's party approaches 40 decibels, 10,000 times louder than the threshold of hearing! The illustration on page 8 shows the decibel levels of these and some other sounds. Prolonged exposure to sound levels louder than 100 decibels can permanently damage your hearing.

High and low sounds

The bottom diagram on this page illustrates *frequency*. It is defined as the number of wave crests or troughs that pass any point in one

Units and properties of sound

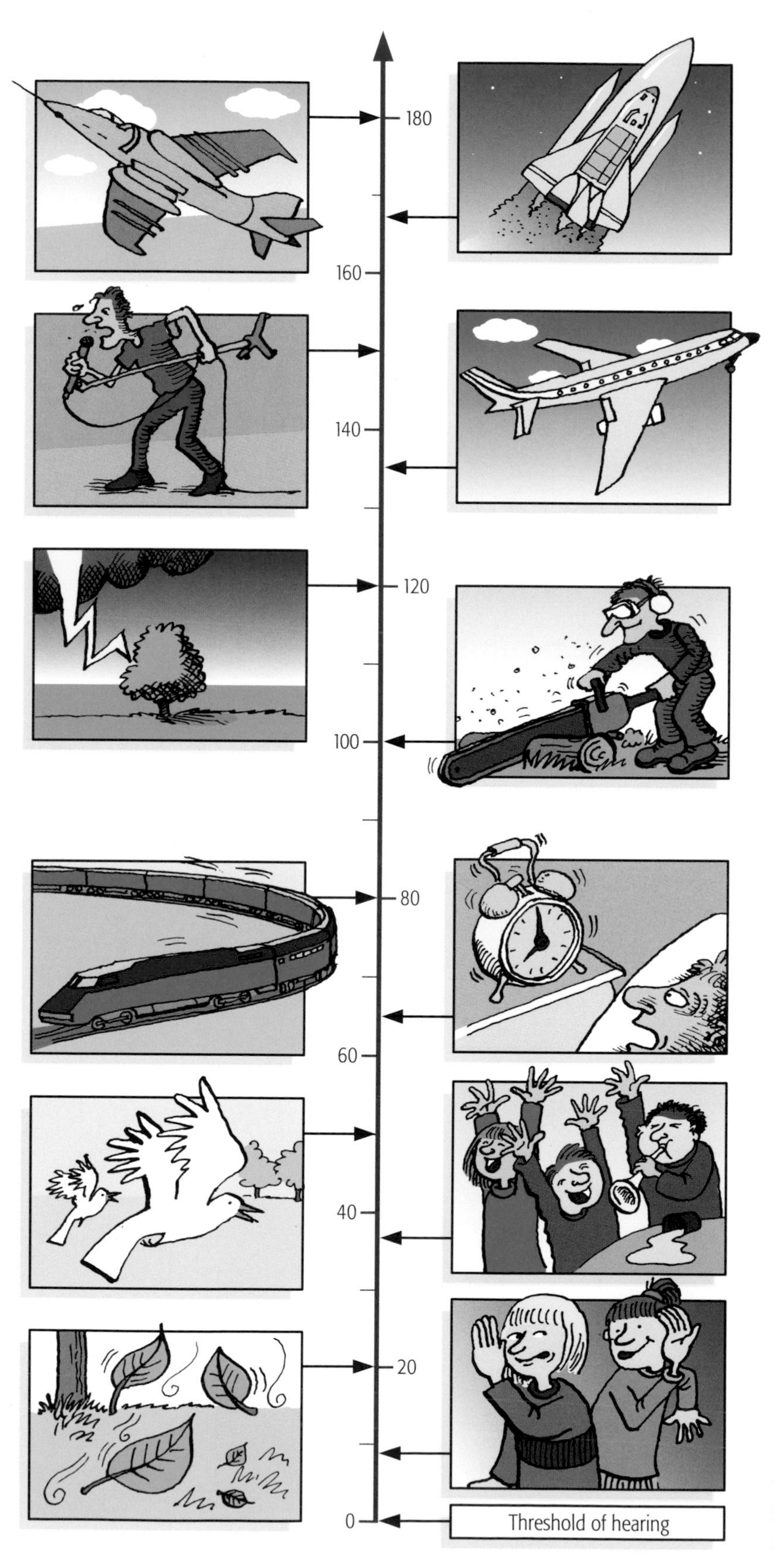

second. The unit of frequency is the hertz, named for the German scientist Heinrich Hertz (1857–94). One crest per second corresponds to a frequency of 1 hertz (symbol Hz). Sounds we can hear have frequencies of between about 20 and 20,000 hertz. As we will see later in this book, some animals can hear sounds at lower and higher frequencies than these.

The larger the frequency of a sound, the higher its pitch. Whistles and screaming babies make high-pitched, high-frequency sounds—as high as 15,000 hertz. A sound with a low frequency has a low pitch. Tubas and ships' foghorns make low-pitched, low-frequency sounds of around 50 hertz.

The speed of sound

If you watch someone using a hammer about 50 meters (45 yards) away, you will see the hammer fall a fraction of a second before you hear its sound. The sound will have taken a certain amount of time to reach your ears, so it must have traveled through the air at a certain speed. In dry air sound has a speed of about 330 meters per second (m/s), equivalent to just over 1,000 feet per second (f.p.s.). So it takes about 3 seconds for sound to travel 1 kilometer, or 5 seconds to travel 1 mile.

You can use these numbers to figure out the distance to a thunderstorm. When you see the lightning flash, start counting in seconds. Stop counting when you hear the thunder. Divide the total number of seconds by 3, and the result is roughly the distance to the storm in kilometers. (Dividing by 5 gives the distance in miles.)

◀ **Sound levels** are measured on the decibel (dB) scale, from less than 10 dB for a whisper to a deafening 180 dB for a military jet on full power.

◀ **Too much noise** is a form of pollution. Here an engineer at an airport is using a decibel meter to measure the noise made by airplanes.

Sound travels faster in materials that are denser than air. For example, it travels at 1,500 m/s (about 4,900 f.p.s.) in water, about 5,000 m/s (16,500 f.p.s.) in metal, and up to 6,000 m/s (19,700 f.p.s.) in rock. So when sound passes from one material into another, its speed changes. When it passes from air into water, for example, its wavelength changes because it speeds up. The musical note A (above middle C) has a wavelength of about 70 centimeters. In water that increases to nearly 3.5 meters. The frequency (pitch) remains the same—the wave just travels faster.

For very high speeds scientists use another unit called the Mach number. It is defined as the speed of an object divided by the speed of sound in the same medium. Because it is a ratio, the Mach number has no units. The speed of sound is called Mach 1; Mach 2 is twice the speed of sound. The first airplane to "break the sound barrier" by exceeding Mach 1 was the U.S. Bell X-1, piloted by Chuck Yeager in 1947. The system is named for the Austrian physicist Ernst Mach (1836–1916), who was the first to photograph the shock waves in front of high-velocity bullets.

Alexander Graham Bell

Alexander Graham Bell gave his name to the bel, a unit of sound intensity, used today in the form of the decibel (dB). However, Bell is best known as the inventor of the telephone. He was born in 1847 in Edinburgh, Scotland. From 1868 he worked in London with his father, a speech therapist; but when his father died in 1870, he emigrated first to Canada and then to the United States. Bell became a professor at Boston University in 1873. His interest in human speech—and deafness—led him to experiment with ways of transmitting sound. He invented the telephone in 1876 and founded the Bell Telephone Company a year later. Bell's later work had to do with airplanes. He died in 1922.

Detecting sound

Humans and all other animals with ears have their own built-in sound detectors. A microphone imitates the action of the ear. Like ears, some microphones are much more sensitive than others. The mouthpiece of a telephone contains the most familiar type of microphone.

Once they have entered the human ear, sound waves make the eardrum vibrate. The eardrum is a membrane, and its vibrations are amplified and converted into nerve impulses that travel to the brain. There, the signals are interpreted as sounds. A microphone is an electromechanical copy of the human ear. All microphones have a vibrating membrane, usually called a diaphragm. Different types of microphone have different ways of converting the diaphragm's vibrations into electrical signals. These signals are amplified and sent to a recorder or loudspeaker, ready to be heard as sounds. In scientific terms a microphone is a kind of transducer—the name for any device that converts one kind of energy into another kind.

Microphones have many applications. The commonest use is in telephones and sound recording. Other devices that incorporate microphones are hearing aids, public-address systems, and dictating machines. Microphones can be made very small for use in hearing aids or as "hands-free" microphones in headsets used by air traffic controllers and people who work in call centers.

Moving-coil microphone

In a moving-coil microphone, incoming sound waves vibrate a diaphragm made of plastic or metal foil. This causes a coil of wire attached to the diaphragm to move rapidly back and forth in the field of a magnet. The movement makes an electric current flow in the coil, and the current varies in step with the sound waves. This varying current is an electrical version of the varying sound waves reaching the microphone. The electrical signal has the same frequency and amplitude as the sound waves. Moving-coil microphones are quite robust and are often used by performers in theaters and concert halls.

◄**A small, lightweight microphone** can be coupled to a headset for listening as well as talking. This person works for a bank at a call center, dealing with queries phoned in by customers.

►**A moving-coil microphone** has a coil of wire that vibrates in a magnetic field to produce an electric current that varies in step with the sound waves.

Ribbon microphone

One of the earliest types of microphone, used in the first days of radio broadcasting, was the ribbon microphone. It has a thin piece of corrugated aluminum foil (the ribbon) attached to the diaphragm. The ribbon is inside the field of a strong magnet. When sound waves vibrate the diaphragm, they also vibrate the ribbon, and that movement makes an electric current flow in the ribbon. As in the moving-coil microphone, this current varies in step with the sound waves.

A ribbon microphone is highly directional—it responds only to sounds coming from a particular direction, which is why it is still much used in radio and TV broadcasting. Like the moving-coil microphone, it is an example of a what is called a dynamic microphone.

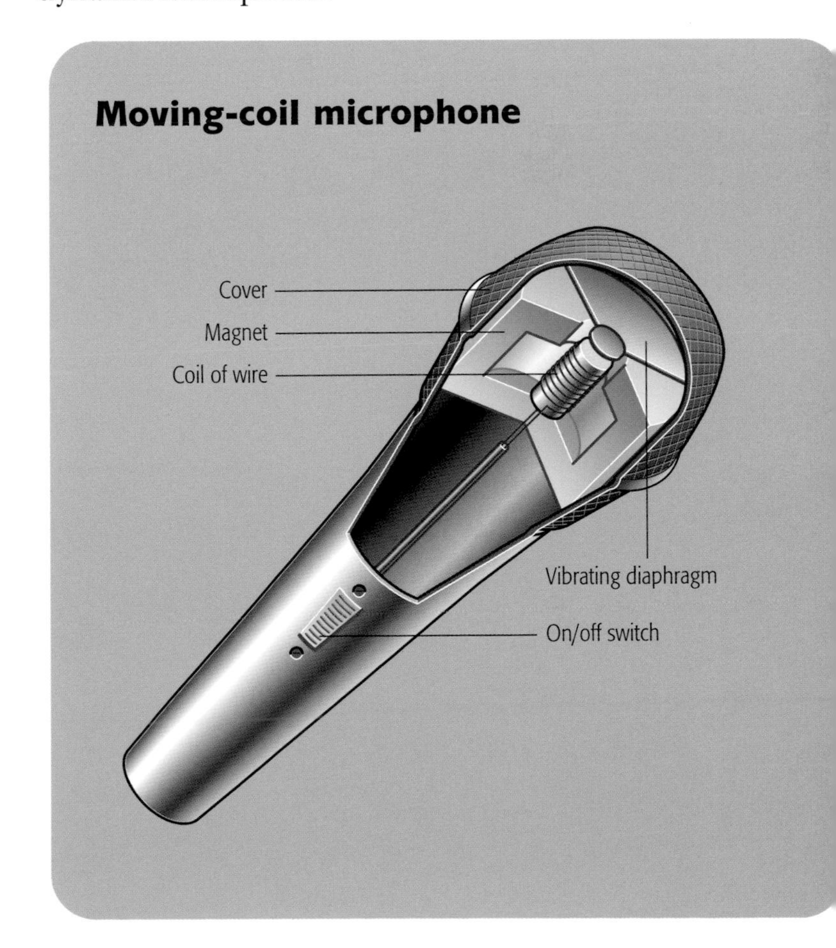

FOR MORE ON DETECTING SOUND SEE *MODERN SOUND RECORDING* **7**:16; *REPRODUCING SOUND* **7**:20; *UNDERWATER ECHOES* **7**:28

Condenser microphone

A condenser microphone also has a diaphragm that vibrates when it is struck by sound waves. Mounted in back of the diaphragm is a fixed metal plate. This arrangement—the diaphragm, the metal plate, and the air gap between them—forms an electrical device called a condenser. The capacitance of the condenser—the amount of electric charge it can store—depends on the width of the air gap. The charge comes from a battery that is connected with the condenser in an electrical circuit.

When sound waves strike the diaphragm, it vibrates. It first moves fractionally closer to the fixed plate, causing the capacitance to increase. It then moves fractionally away from the fixed plate, and the capacitance decreases. As a result, the electrical output from the condenser varies in step with the sound waves. A condenser microphone can be made very small and is the type favored for hearing aids. Recording studios use condenser microphones because, although they are expensive, they provide very good sound reproduction.

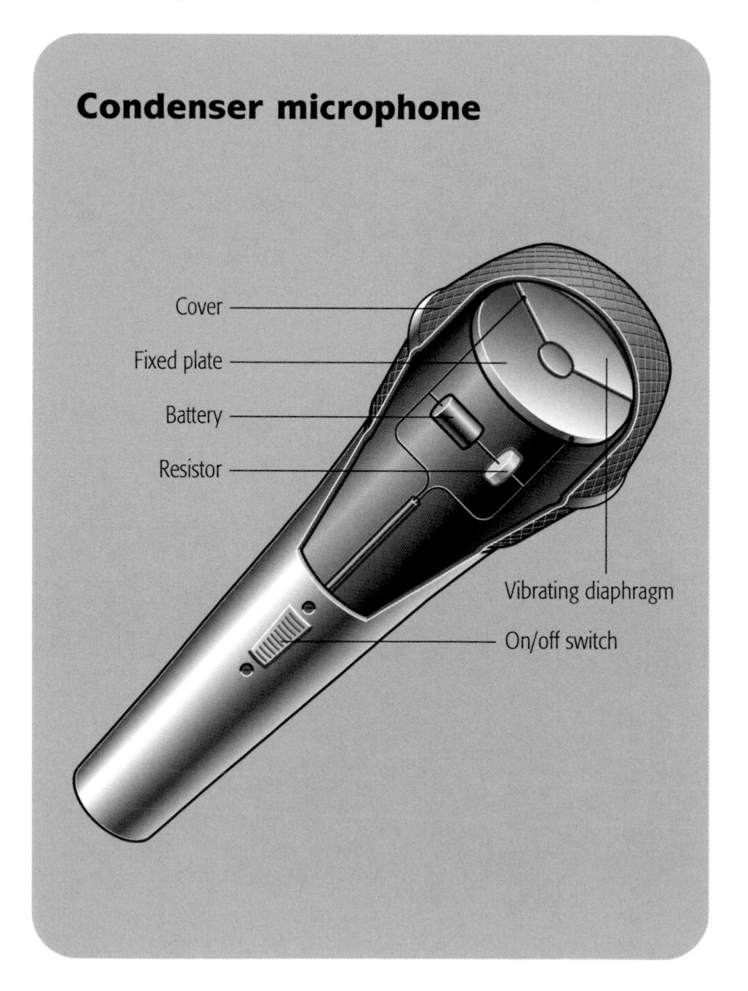

◀**A condenser microphone's diaphragm** and the fixed plate behind it together make up an electrical condenser, which forms part of a circuit that also includes a small battery. Vibration of the diaphragm makes the capacitance of the condenser vary in step with the sound waves that vibrate the diaphragm. This results in a varying current in the circuit that becomes the output of the microphone.

Crystal microphone

A crystal microphone makes use of a phenomenon called piezoelectricity: Certain crystals produce an electric voltage when they are squeezed. The French physicist Pierre Curie discovered the effect in 1880. The microphone has a crystal of quartz sandwiched between two metal plates, which act as electrodes. One end of the crystal is fixed, and the other end is attached to the diaphragm.

When sound waves vibrate the diaphragm, it alternately squeezes and releases the crystal. This makes the crystal give out a varying voltage, which is picked up by the electrodes. Once again, the voltage varies in step with the sound waves and provides the electrical output of the microphone. People conducting outdoor interviews for TV news broadcasts commonly use crystal microphones.

▼A famous foursome face a battery of microphones as they answer questions from reporters. The Beatles also made good use of microphones on stage and when they were recording their music.

Carbon microphone

The carbon microphone was the first type of microphone to be invented, by the British-born American engineer David Hughes, in 1878. It was much improved by Alexander Graham Bell in the following years, and it became the standard microphone for telephones for more than a century.

The chief components of a carbon microphone are a cup-shaped container of granules (very small pieces) of carbon and a diaphragm. A metal plate attached to the diaphragm presses down on the granules, which—together with a battery—are connected together in an electric circuit. Pure carbon conducts electricity fairly well, and how much the granules conduct depends on how closely they are packed together. Packed tightly, the granules have a lower electrical resistance and so conduct more electricity than when they are loose.

When sound waves make the diaphragm vibrate, it presses on the metal plate. This action alternately squeezes the granules and releases them, changing their resistance, so the voltage in the circuit varies in step with the sound waves. A carbon microphone will capture only a narrow range of frequencies, but is good enough for reproducing speech on a telephone.

David Hughes

The American inventor David Edward Hughes was born in London, England, in 1831. He emigrated with his parents to the United States in 1838. He was raised in Virginia and took up music, and from 1850 to 1853 he was professor of music at St. Joseph's College in Bardston, Kentucky. In the 1850s he began to experiment with printing telegraphs, which he patented in 1856. He sold most of his machines in Europe, and his designs remained in use until the 1930s.

Hughes also built electromagnets and in 1878 made his first carbon microphone—a term he coined himself. A year later he gave a practical demonstration of radio waves, but did not publish his findings until 1899. He died a year later in London, where he had lived and worked since 1877, and left his considerable fortune to London hospitals.

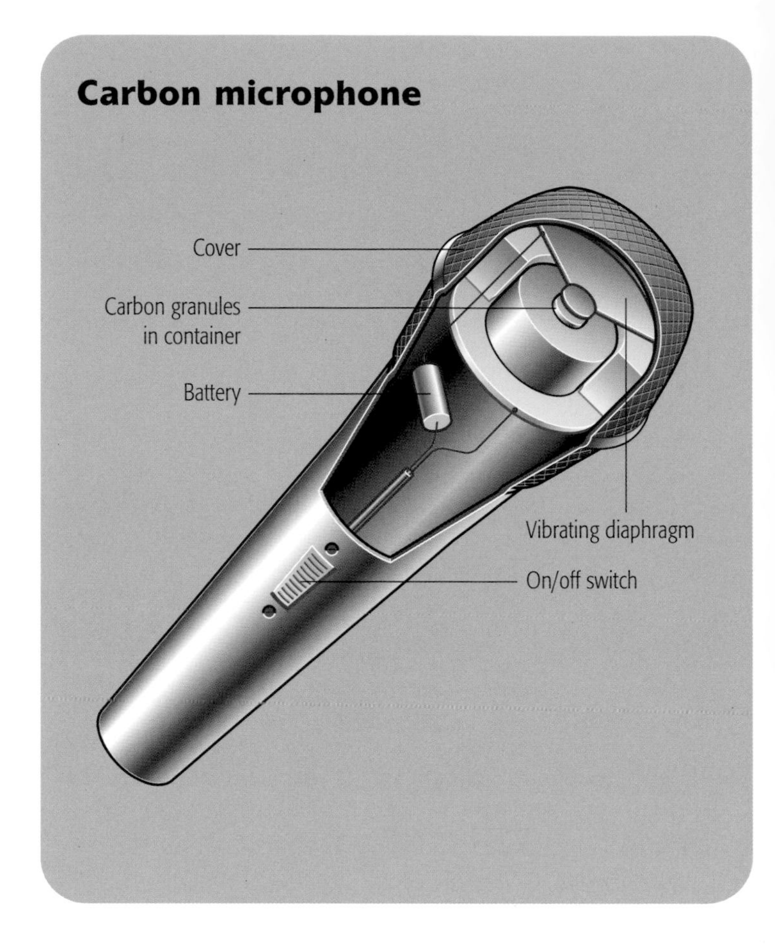

▲ **In a carbon microphone** the vibration of the diaphragm presses on carbon granules, changing their electrical resistance. The changes in resistance lead to changes in the output voltage of the microphone.

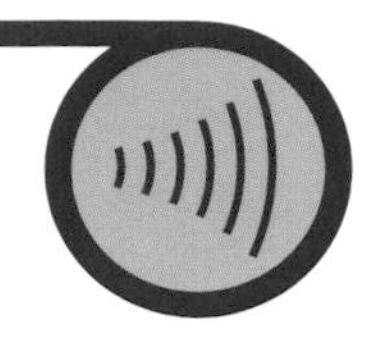

Relaying sound signals

As we have seen, the purpose of any type of microphone is to convert sounds into a varying electric current or voltage, which can pass along wires. Telephone calls were the first electric sound signals to be transmitted over fairly long distances, along wires. But engineers soon found that there was a limit to how far the signals would go before their quality deteriorated and the caller's voice became so distorted that it could not be understood. They also suffered from interference from any nearby electrical machinery.

Then, beginning in 1936, telephone engineers began to use coaxial cables. A modern coaxial cable has a central wire conductor surrounded by plastic insulation. It in turn is surrounded by a tube of copper or copper braid. This outer layer of copper shields the inner wire from interference. It is possible to bundle together up to twenty cables to make a trunk cable that can carry 130,000 high-frequency phone messages at the same time.

Coaxial telephone cables could be laid on the seabed to connect continents. The first, between Newfoundland and Scotland, was installed in 1956. Even so, it had to have amplifiers (called repeaters) built into it every few kilometers to boost the signals. The introduction of digital signals in the 1970s allowed coaxial cables to carry 40,000 two-way conversations on ten cables bundled together. In recent years many coaxial cables have been replaced by fiber-optic cables for landlines and by microwave satellite links for intercontinental telephone calls. A single fiber-optic cable can carry 50,000 two-way conversations at once. Satellite microwave links can each carry several TV channels as well as 33,000 telephone calls.

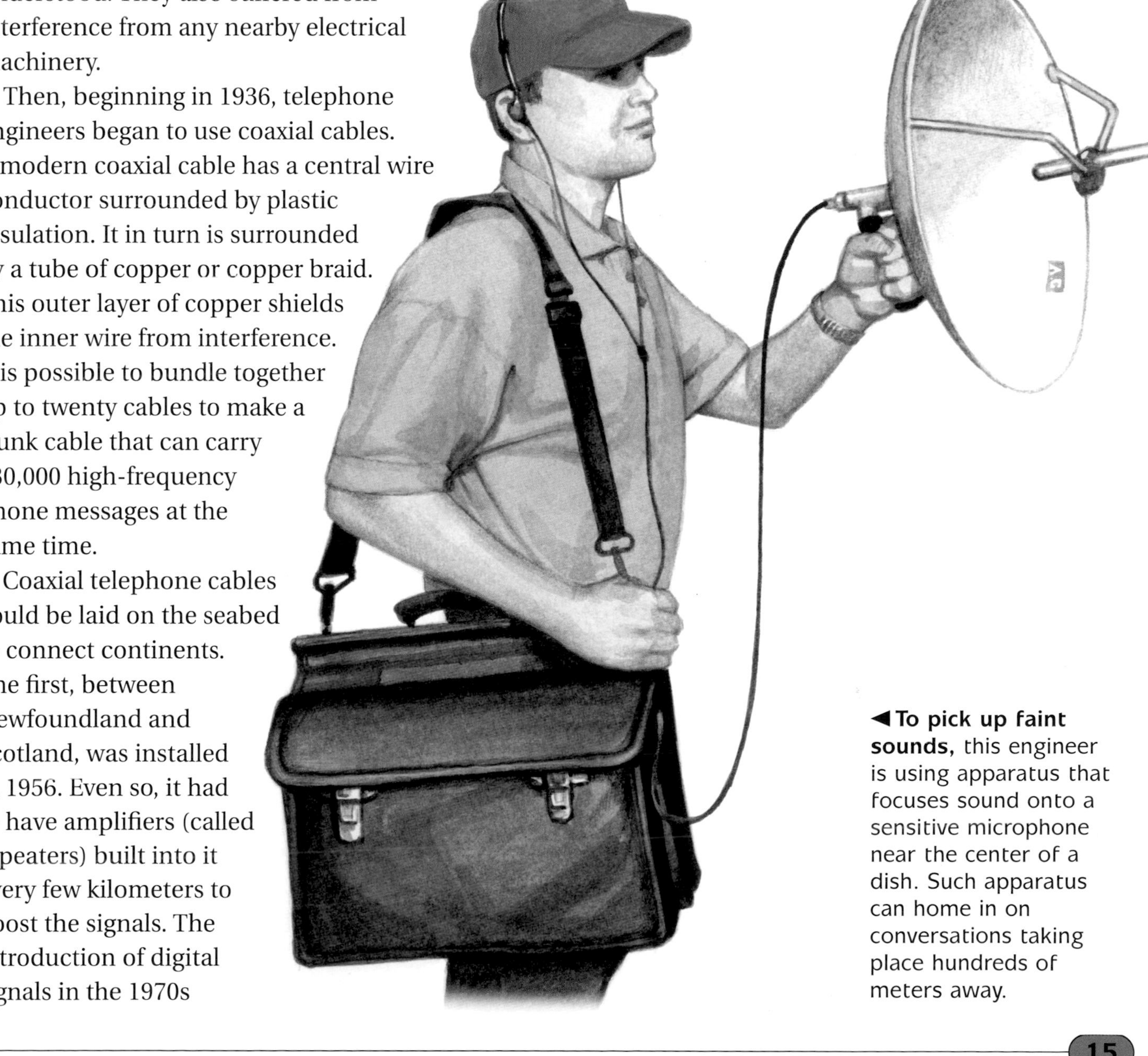

◀ **To pick up faint sounds,** this engineer is using apparatus that focuses sound onto a sensitive microphone near the center of a dish. Such apparatus can home in on conversations taking place hundreds of meters away.

Modern sound recording

Sound recording, in its first stage at least, uses magnetic tape. Prerecorded tapes are easy to carry around and can be played back almost anywhere. Digital techniques produce the best recordings on tape or on compact disks, which are played on a CD player incorporating a laser.

We have seen on the previous pages how a microphone converts sounds into a series of varying electrical signals. The purpose of modern sound recording is to preserve these signals so that they can be played back at any time. As a first stage, nearly all sound-recording methods use a tape recorder, illustrated opposite.

Initially, a microphone picks up the incoming sounds and passes them, in the form of varying electrical signals, to an amplifier. The amplifier boosts the voltage of the signals before it sends them to the record head of the tape recorder. This head is essentially an electromagnet. The varying signal into the head makes the electromagnet produce a varying electric field. An electric motor (not shown) rotates the takeup spool and feeds magnetic tape past the record head.

Magnetic tape is a ribbon of thin plastic coated on one side with a layer containing small particles of a magnetizable material, such as iron or iron oxide. On unrecorded "blank" tape the iron oxide particles are arranged in a completely random manner within the layer. When the tape passes the record head, the magnetic field magnetizes the particles and lines them up in a regular pattern. This effectively "freezes" the magnetizing field on the tape.

◀ **The clapper board** in a film studio makes a bang that is used to synchronize the sound with the pictures on a movie film.

Tape recording

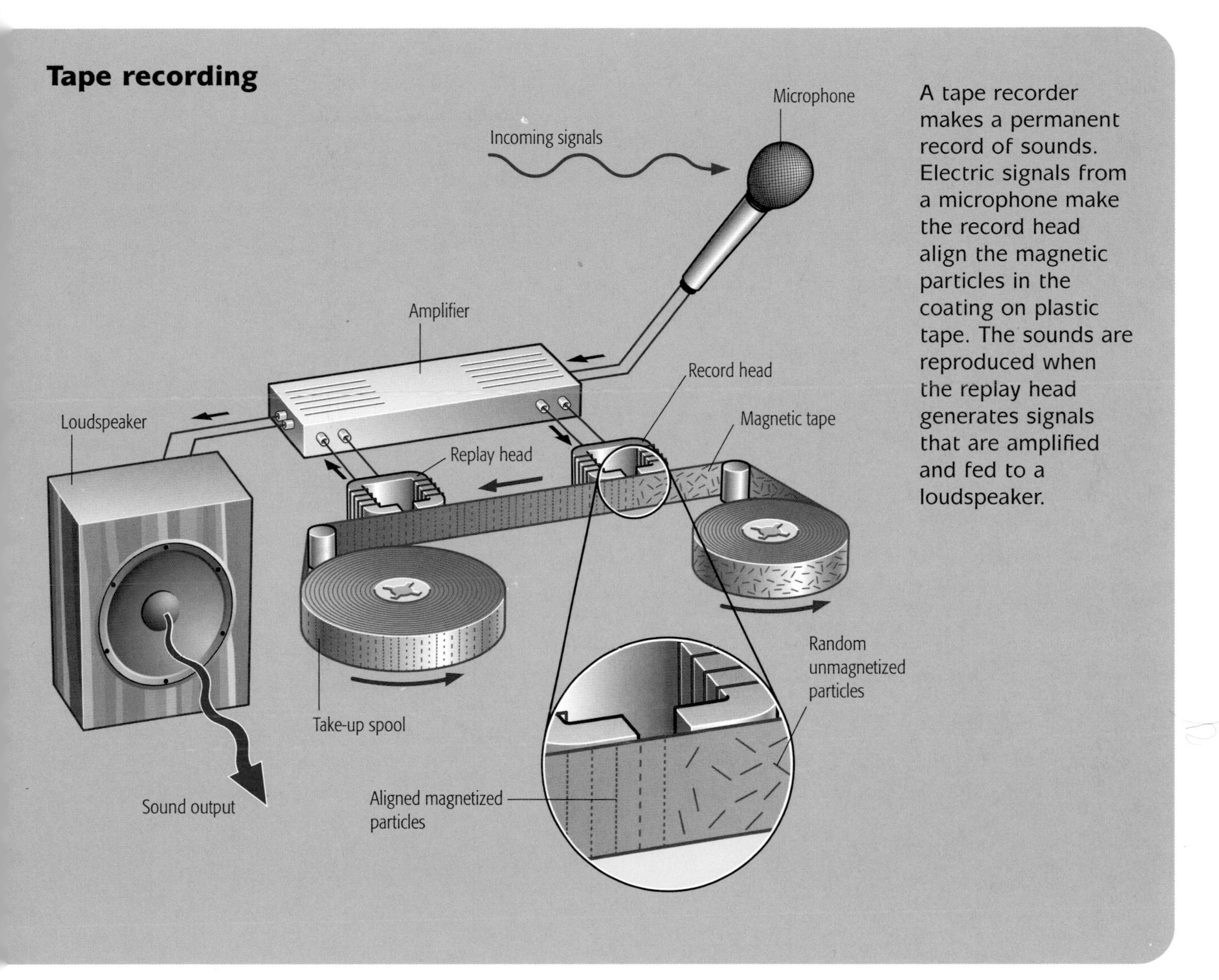

A tape recorder makes a permanent record of sounds. Electric signals from a microphone make the record head align the magnetic particles in the coating on plastic tape. The sounds are reproduced when the replay head generates signals that are amplified and fed to a loudspeaker.

On playback the prerecorded tape passes in the same direction as before in front of a replay head. This time the miniature magnets on the tape create tiny varying voltages in the wires of the replay head—a bit like an electromagnet working in reverse. The small voltages pass to the amplifier, where they are made much larger before going to a loudspeaker. As described on page 21, the loudspeaker converts the varying electric signals back into sounds. The sequence of steps from recording to playback is now complete.

Analog to digital

Because the varying electrical signals entering and leaving the tape recorder are counterparts of the changing sound signals reaching the microphone, they are called analog signals. The signals are not exact replicas of the sounds, however, so analog recordings suffer from some distortion of sound. One way around this problem is to convert the analog signal into a digital signal before recording it.

The key to analog-to-digital conversion is a technique known as sampling in which the

FOR MORE ON MODERN SOUND RECORDING SEE *USING MICROWAVES* **6**:*32*; *DETECTING AND RELAYING SOUND* **7**:*10*; *REPRODUCING SOUND* **7**:*20*

amplitude (height) of the varying analog signal is measured many times each second—a modern audio CD has a sampling rate of 44,100 times a second (44.1 kilohertz). In the illustration below, the red line represents the varying analog audio signal. To make it easier to see what is happening in this illustration, the signal is sampled just ten times, giving amplitudes beginning at 31 on the left and ending at 25 on the right. An actual converter can recognize more than 65,000 different levels of amplitude.

Next, electronic circuits convert the sampled amplitudes into binary numbers. They are numbers that have only two digits, 1 and 0 (the decimal numbers that we normally use have ten digits: 1 to 9 and 0). Binary digits are chosen because they can be made to correspond to "off" and "on" pulses of electric current.

The table at the right of the illustration lists the 8-bit binary numbers that correspond to the ten sampled amplitudes. Notice that they consist only of 1's and 0's, and that they can be strung together to give the continuous digital signal shown at the bottom of the illustration. What has happened is that the continuously varying analog signal (from a microphone) has been converted into a series of "off" and "on" pulses of electricity—the digital signal. This signal can then be recorded on magnetic tape or a compact disk (CD).

▼ **In digital recording** an analog signal (a varying voltage corresponding to a series of sounds) is sampled regularly, and each value is converted into a series of "on" and "off" pulses that together form the digital signal. The greater the sampling rate, the better the quality of the recorded sound.

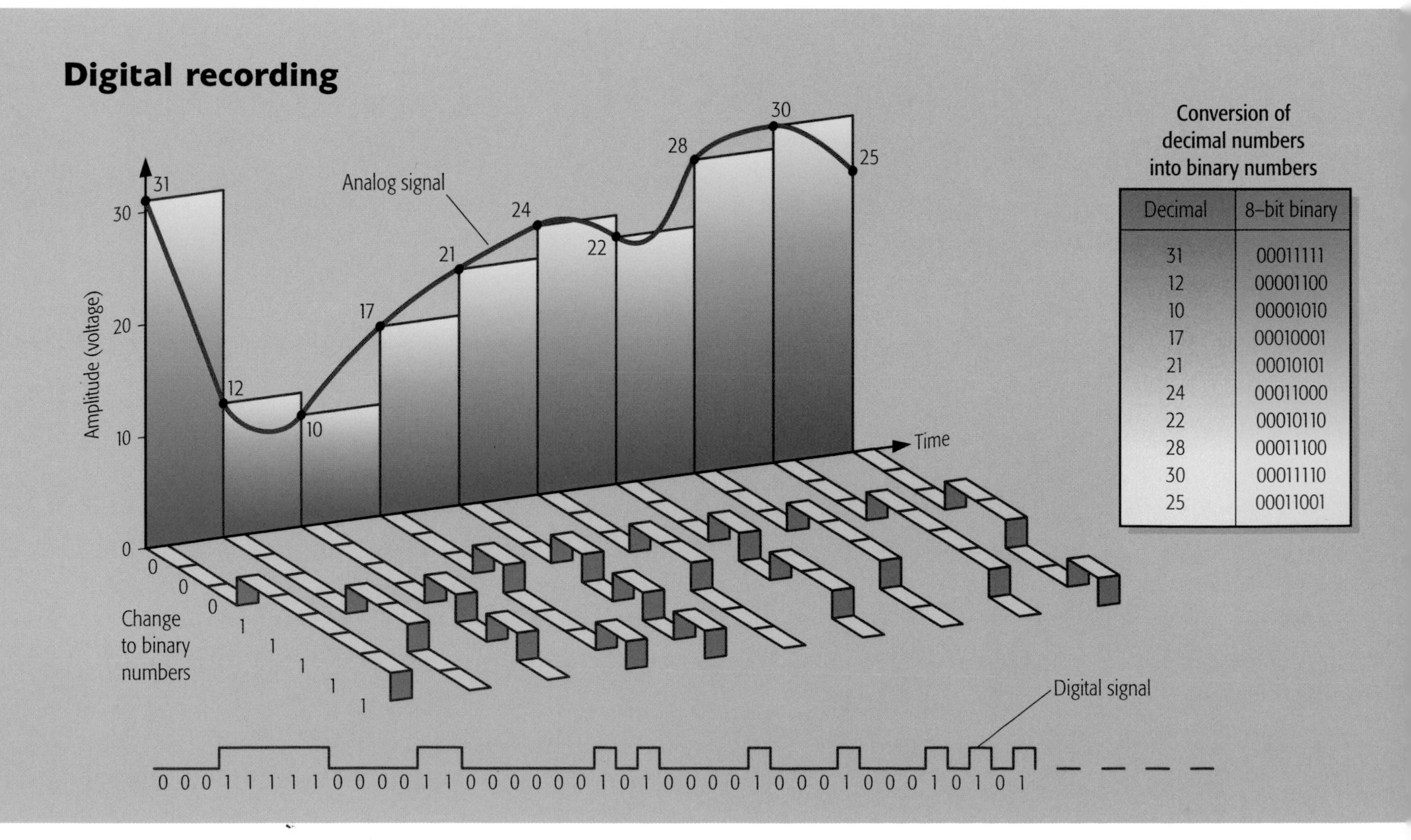

Decimal	8-bit binary
31	00011111
12	00001100
10	00001010
17	00010001
21	00010101
24	00011000
22	00010110
28	00011100
30	00011110
25	00011001

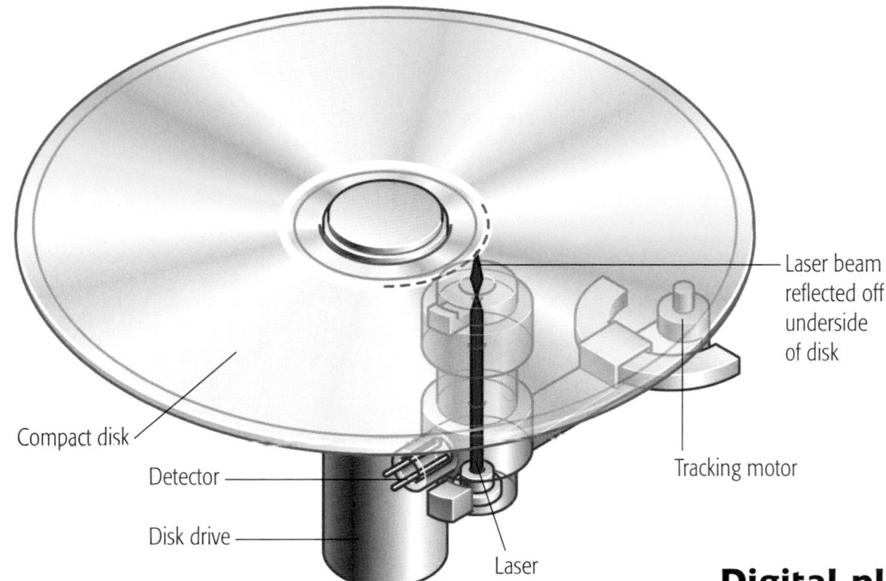

◀ **In a compact disk player** a laser beam is reflected off the underside of a CD. The reflections, which vary according to the series of minute pits and flats on the disk, are converted into a sequence of varying electric signals that are amplified to produce sounds in a loudspeaker.

Movie soundtrack

A method of recording sound on film as an optical soundtrack was invented in 1923 by the American electrical engineer Lee de Forest. In his system the varying electric current from a microphone is made to vary the light output from an electric lamp. This light shines on a track along the side of the movie film, and the variations in width or density of the exposed track correspond to variations in the pitch and intensity of the sound.

On playback light from a lamp in the movie projector shines onto the soundtrack along the edge of the film. The light that passes through, which varies with the width of the track, is picked up by a photoelectric cell. It is a device that produces an electric current when light falls on it. The output of the photoelectric cell varies in step with the sound signal and can be amplified to reproduce the sounds. Because the soundtrack is part of the film, the sound is synchronized exactly with the action of the movie.

Digital playback

Playing back a digital audio tape (DAT) is much the same as playing back an analog tape (see page 17), except that the output signal is digital and has to be processed by a digital amplifier. Playing back a digital disk, however, is very different from playing an analog vinyl disk on a phonograph.

A digital disk, usually known as a compact disk (CD), is a 12-centimeter (4.5-in.) aluminum disk encased in transparent plastic. A series of pits are etched into the aluminum in a long spiral "groove" (on the playing surface, the underside). The leading edge of a pit represents the digit 1 in the digital sequence; a long pit stands for a series of 1's. The areas between pits, called flats, represent 0's.

As the disk spins around in the CD player, lenses focus a laser beam onto the shiny underside of the disk. When the laser light strikes a flat (corresponding to a 0), it reflects off the disk and onto a detector. When the light strikes a pit (corresponding to a 1), the light is scattered and does not reach the detector. In this way the output signal from the detector corresponds to the 0's and 1's of the original digital signal. An amplifier and loudspeaker system then change the digital signal into an exact replica of the original recorded sound.

Reproducing sound

A loudspeaker converts electrical signals into sounds. The signals may come directly from a microphone, from a live radio or TV broadcast, or from a recording medium such as magnetic tape or a compact disk. The design of the speaker is chosen to suit the pitch of the sounds.

Loudspeakers work like microphones in reverse—they convert electrical signals into sounds. And like microphones, loudspeakers are transducers, converting one kind of energy into another kind. They have various applications: in home sound reproduction equipment such as cassette recorders, record players, and CD players; in telephone receivers; in public-address systems; and in radios and TV receivers. They are also used at theaters, discos, and rock concerts.

Dynamic speakers

Most regular loudspeakers, generally called just "speakers," are of the dynamic kind—they have moving parts. Like the moving-coil microphone that it resembles (see page 11), it has a coil of fine wire placed between the poles of a powerful circular magnet. One end of the coil is attached to the center of a cone of paper or plastic. The wires from the other end of the coil are connected to terminals leading to the source of the audio-frequency electrical signals, usually an amplifier.

The varying audio signal flowing in the coil makes the coil move back and forth. As it does so, it makes the speaker cone vibrate. It is the vibration of the cone that produces the sound. Large cones, called woofers, vibrate best at low frequencies (deep notes), while small cones, called tweeters, are better at reproducing high frequencies (high notes). For this reason, many stereo systems have two or more loudspeakers to cover the whole audible frequency range, from about 20 or 30 hertz (Hz) to 20,000 Hz, the upper limit of human hearing. An electronic circuit called a crossover network splits the frequencies to suit the individual speakers.

The size and shape of the speaker cabinet, or "housing," can affect the quality of the sound produced. The commonest type is an airtight box lined with sound-absorbent flock or plastic sponge. For better reproduction of low frequencies there may be a hole in the box called a bass-reflex opening, or a tube may connect the inside of the enclosure to the outside. For a large output (volume) a flared tube called an acoustic horn may be attached to the speaker cone. It is the type of speaker favored for public-address systems and other outdoor applications.

◀ **A disco** would not be the same without the barrage of loud sounds generated by the DJ's sound system.

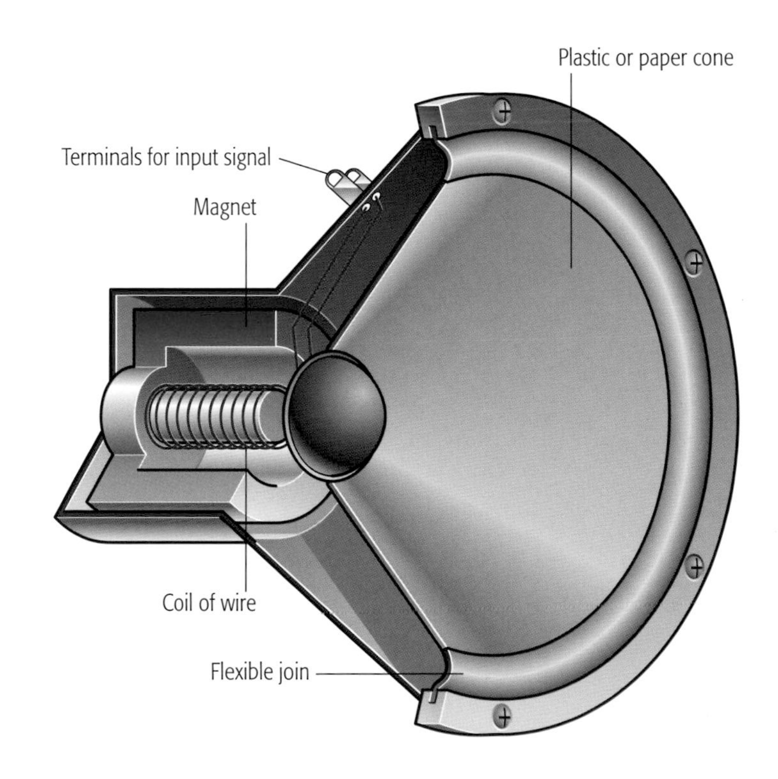

▲ **A loudspeaker** has a coil of wire attached to a cone of paper or plastic. The audio signal flows through the coil, which vibrates in the field of a large magnet. The coil's vibrations are passed on to the speaker cone to produce sounds.

FOR MORE ON REPRODUCING SOUND SEE *USING MICROWAVES* **6**:*32*; *DETECTING AND RELAYING SOUND* **7**:*10*; *MODERN SOUND RECORDING* **7**:*16*

Other kinds of speakers

A powerful loudspeaker and its enclosure tend to be large and take up a lot of room, which is a problem if you have a stereo in a small apartment. The answer is the electrostatic speaker, which can be slim enough to hang on a wall. It consists of a thin sheet of metal held between a pair of parallel screens. The audio signal passes to the screens, causing positive and negative electric charges to form on each side of the metal sheet. This creates an electrostatic force, which makes the sheet vibrate and produce sound waves. Electrostatic speakers handle low frequencies better than high frequencies.

An earphone is a small loudspeaker that is held close to the ear or, if small enough, placed inside the ear canal (as with a hearing aid). The speaker in a telephone handset is a type of earphone. A pair of earphones mounted on a band worn on the head are usually called headphones. If they also incorporate a small microphone, they are called a headset (see the photograph on page 10). Earphones are usually worn where there is a lot of background noise, as on the flight deck of an airliner or in factories where there are several machine tools working at the same time.

Michael Faraday

Michael Faraday was an English physicist who was born in 1791 in the southern town of Newington. His father was a blacksmith. He had little formal education but went on to become one of the greatest chemists and physicists of his time.

His many discoveries include the organic substance benzene, the laws of electrolysis, the dynamo, and the transformer. These last two make use of the relationship between magnetism and electricity, which was later to become so important to the development of microphones and loudspeakers. Faraday showed that moving a coil of wire in a magnetic field produces an electric current (as in a moving-coil microphone) and the opposite—that passing a current through a coil in a magnetic field causes the coil to move (as in a loudspeaker).

Throughout his life (he died in 1867) Farraday worked to bring science to ordinary people, especially children. The SI unit of capacitance, the farad (symbol F), was named in his honor.

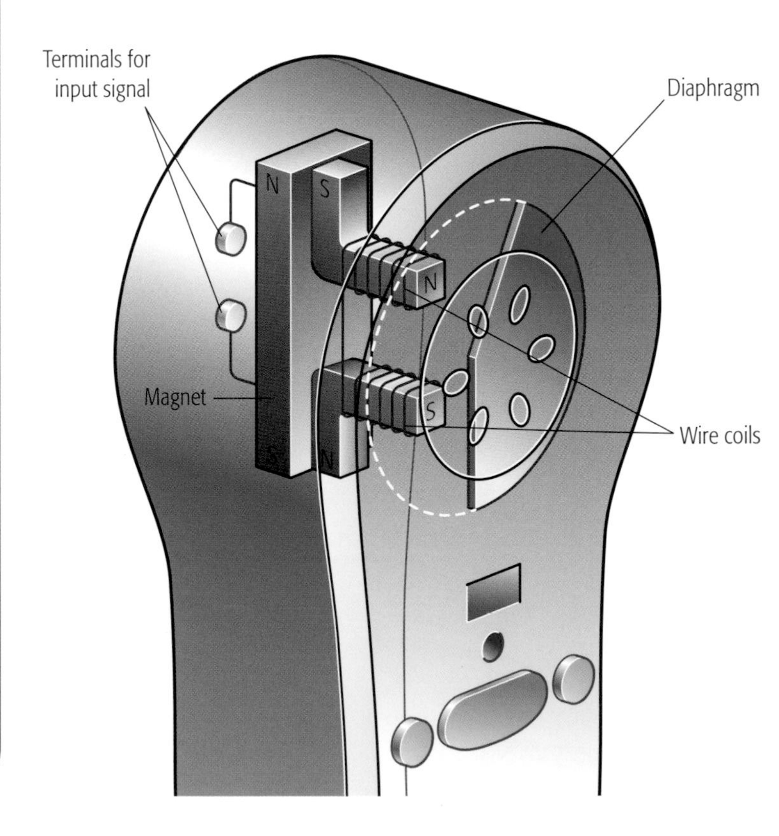

Synthetic sound

So far we have looked at devices that reproduce real sounds, for example, from a stage performer or from a recording. But it is also possible to use electronics to create original, never-before-heard sounds. Electric guitars, electronic organs, and various kinds of synthesizers, for example, can produce electronic music.

An electric guitar has a series of pickups under its metal strings. When the guitarist plucks a string, it vibrates, and the pickups convert the vibrations into electrical signals that can drive an amplifier and loudspeaker or be recorded. A fuzzy sound or an echo can be added to the notes by an effects processor in the guitar amplifier. Even more variation can be provided by linking an electric guitar or other electronic instrument through a computer.

In the 1950s the American electronic engineer Robert Moog developed the Moog synthesizer. His original instrument could play only one note at a time, but it was entirely electronic—no strings—and was played using a keyboard.

Modern synthesizers can accurately reproduce the sound of any musical instrument and provide a digital output for recordings. They can incorporate a drum machine for percussion effects, and some synthesizers are used to create special sound effects for TV and movies.

◀ **A telephone earpiece** contains a simple loudspeaker. In this older type the incoming speech signal converts two L-shaped pieces of iron into electromagnets, which vibrate a metal diaphragm to produce sounds.

▲ **The keyboard player** in a rock band is the ultimate sound producer. The keyboards are linked to computers and synthesizers that can produce almost any kind of sound.

Ultrasound — too high to hear

The pitch (frequency) of ultrasonic sound is too high for the human ear, though some animals, such as bats and dogs, can hear it. Scientists have found various uses for ultrasound, including ultrasonic scanners employed in medicine and industry to "see" inside people or objects.

The upper limit of human hearing is a frequency of about 20,000 hertz (Hz). Higher frequencies than this are called ultrasound. Its most familiar use is for making ultrasound scans of body tissues. This painless procedure is safer than using x-rays, which, because they are a form of ionizing radiation, may harm delicate tissues such as those of a baby developing in its mother's womb. The process is illustrated on these pages.

The scanner head contains many ultrasonic transducers that convert a varying voltage into pulses of ultrasound at a frequency of about 30,000 Hz. The transducers consist of piezoelectric crystals, which vibrate at ultrasonic frequencies under the influence of the voltage. A computer controls which transducers are working at any given moment, and by switching them on and off in turn, it makes the beam of ultrasound scan sideways.

Like ordinary sound, ultrasound travels at different speeds in different materials, depending on their density. Also, ultrasound waves are reflected when they pass from one material to another. These echoes bounce back to the scanner, which detects them and converts them into small voltages that are passed to the computer. A special

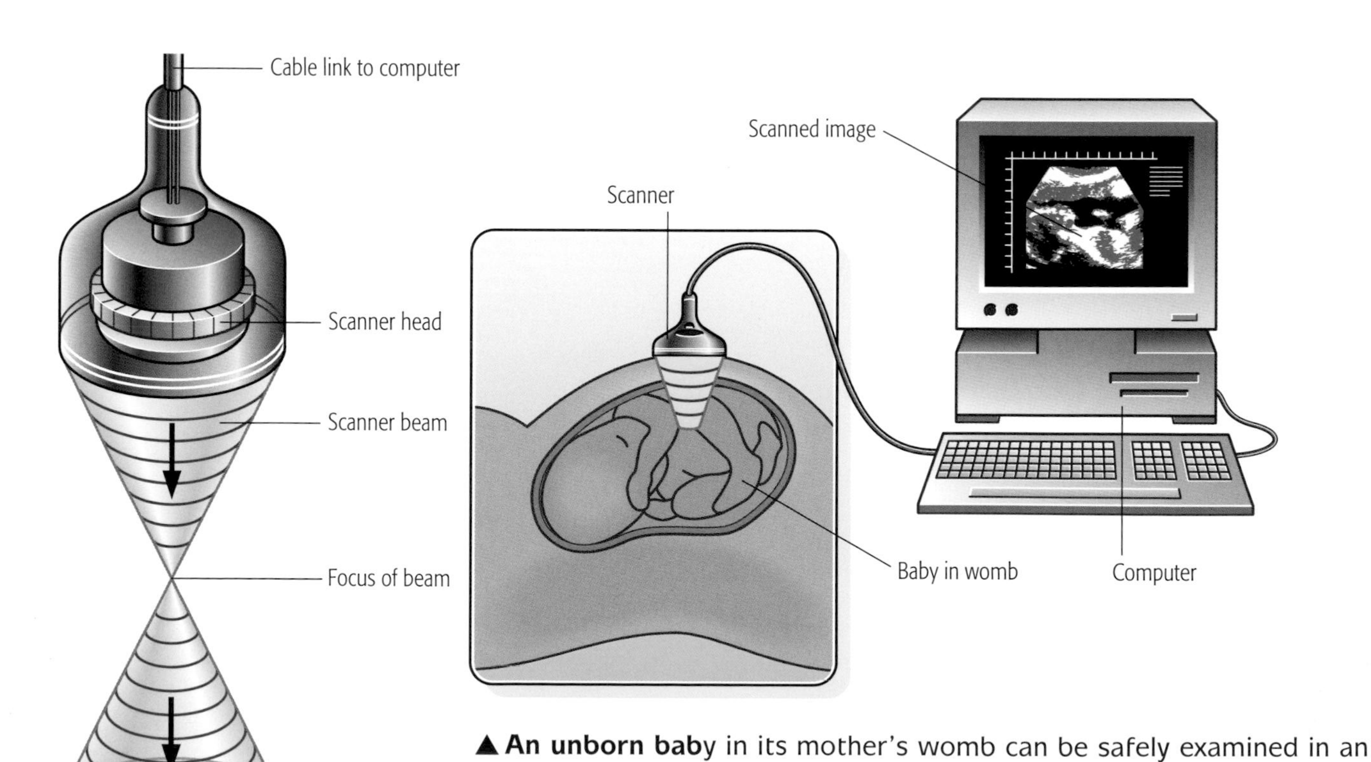

▲ **An unborn baby** in its mother's womb can be safely examined in an ultrasound scan. The scanning head (left) produces a cone of ultrasound pulses. Reflections (echoes) from the baby's tissues are converted into electronic signals that a computer forms into a picture.

Ultrasound scanning

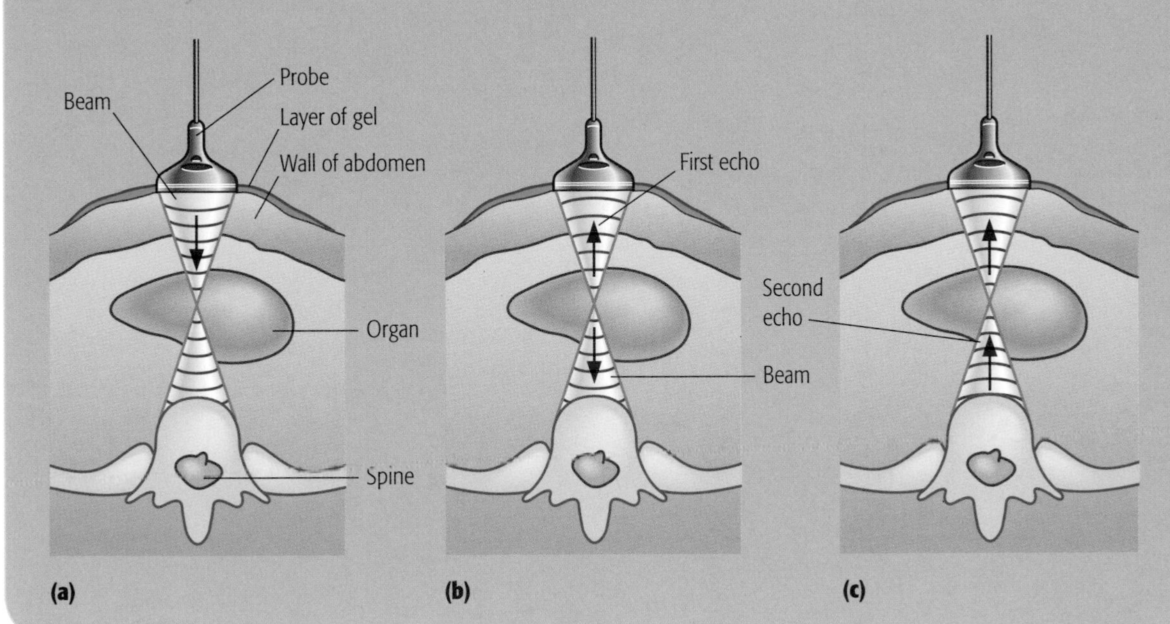

Different tissues reflect ultrasound in different ways. These diagrams represent sections through a person's abdomen. The ultrasound beam passes through the abdominal wall (**a**) and is reflected by an internal organ (**b**). It then passes through the organ before being reflected by other tissues farther away (**c**).

jellylike substance called a gel is applied between the scanner and the person's skin to prevent unwanted echoes from the skin surface. From the strength of the echoes and the time they take to be reflected off different tissues, the computer builds up an image of the body tissues and displays it on the computer screen.

Other medical uses

A woman generally has an ultrasound scan to confirm she is pregnant about 7 weeks into her pregnancy and another at about 20 weeks to check on the progress of the baby. Other medical uses of ultrasonics include scanning for tumors and kidney stones and to detect possible defects in the heart. Ultrasound is also used to treat various brain disorders, and powerful ultrasonic beams can be used to break up kidney stones painlessly. By applying the principles of tomography (see the section on CAT scanning in Volume 6 of *Science Tools*), the computer can be made to assemble a set of cross-sectional ultrasound scans to produce a three-dimensional "picture" of the brain or other body tissues.

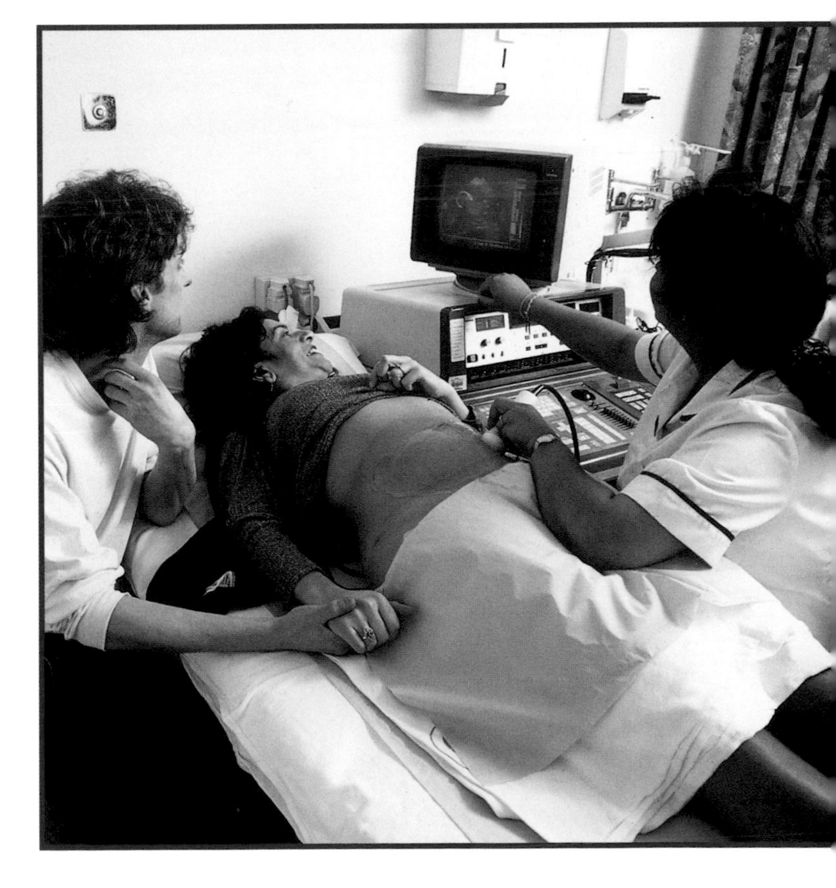

▲ **An ultrasound scan** has become a routine and painless way of examining a pregnant woman. It can check that all is well and provide early warning of any problems.

FOR MORE ON ULTRASOUND SEE *X-RAYS AND GAMMA RAYS* **6**:*26*; *UNDERWATER ECHOES* **7**:*28*; *INFRASOUND–TOO LOW TO HEAR* **7**:*32*

Ultrasound — too high to hear

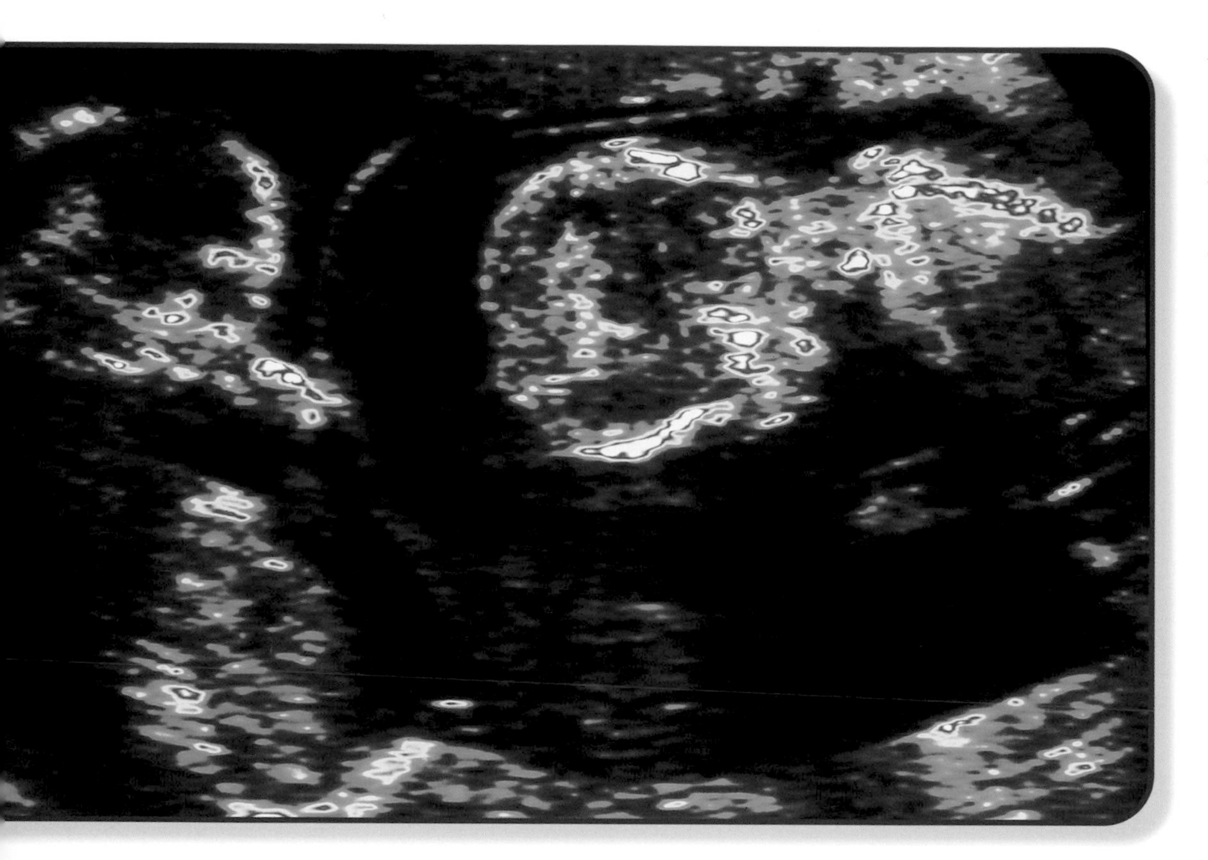

◀ **It's twins!** This ultrasound scan shows that the mother is expecting twin babies. The one on the left has its head uppermost, while the one on the right is lying on its side.

Industrial uses of ultrasound

In one of the chief industrial applications of ultrasound a scanning system similar to that used in medical examinations is used to probe the interior structure of metal objects and detect flaws. High-power signals of 1 to 20 MHz (1–20 megahertz, i.e., 1–20 million hertz) enable operators to detect cracks, fractures, holes within castings, and poorly welded joints. Pipelines for carrying oil or water, which are usually built in sections that are welded together, are examined by an ultrasound scanner mounted on a trolley that travels along the inside of the pipe. A computer analyzes the echoes and displays an image of the metal's structure on a monitor.

High-power ultrasound can be used to shake, or "agitate," liquids very vigorously. This is one way of mixing an oily liquid with a watery one to form an emulsion, and it is often used to make emulsion paint. Other uses of ultrasound include killing bacteria in milk and welding together sheets of plastic (see the illustration opposite). The high-speed vibrations produced by an ultrasonic oscillator can be used to melt the surfaces of two sheets of plastic and weld them together.

Ultrasound in the animal world

Animals are not science tools, but it is interesting that various animals can produce ultrasound. Bats emit ultrasonic pulses and use the returning echoes to locate their prey in the dark and to navigate around obstructions. Members of the whale family also navigate and communicate by using ultrasound. Many insects use ultrasound signals to call for their mates. So what may seem to be a silent night to human ears is probably alive with supersonic squeaks and chirrups. As we shall see on the pages that follow, humans have learned how to copy this natural navigation technique for their own purposes.

► **Rapid up-and-down vibrations** produced by an ultrasonic oscillator can be used to weld plastic sheets together.

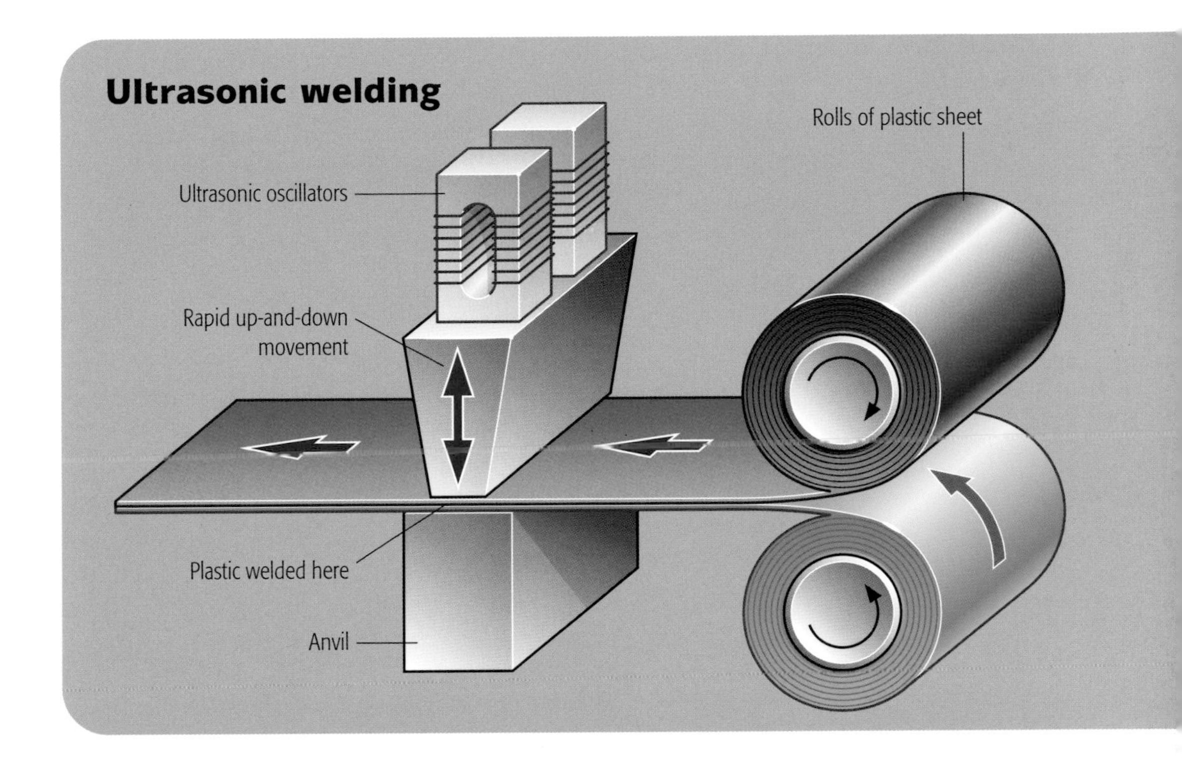

▼ **A bat** is a natural producer of ultrasonic sound signals. Flying often in total darkness, it gives out high-pitched ultrasonic "squeaks" and uses its sensitive ears to pick up any echoes. It then zooms in on the source of the echo, which often makes a tasty meal.

Underwater echoes

At one time sea fish could safely hide near the ocean bottom or in and around shipwrecks. And the wrecks themselves were a danger to shipping. A technique called sonar, originally developed to detect icebergs and submarines, uses pulses of ultrasound to find the fish and the wrecks.

Sometimes the meaning of a scientific word is not quite what you might think. Take the word "sounding." In science its main meaning is "finding the depth of water under a boat." In the past, finding the depth was the job of a crew member known as the leadsman. He had a length of rope with knots tied at every fathom (6 feet, about 1.8 meters) and a heavy lead weight at the end. He would swing the lead and drop the line into the water. The lead weight would sink to the bottom. By counting the knots pulled into the water, he knew how deep the water was. That is why the modern method of doing the same thing is called echo sounding (not because of the sound waves it uses).

An echo sounder works by emitting and detecting pulses of ultrasound. A transmitter on the ship's keel sends the pulses downward. The seabed reflects the pulses back up to the ship, where a special microphone called a hydrophone picks them up. From a measurement of the time it takes for a pulse to leave the ship and its echo to return, the distance to the bottom—the depth of water—can be calculated. So the echo sounder makes soundings by measuring echoes.

◀ **Searching for fish,** a trawler transmits an arc of sonar signals. Echoes reflected by a shoal of fish indicate where the fish are and how far away they are. The ship alters course to trawl its net through the shoal.

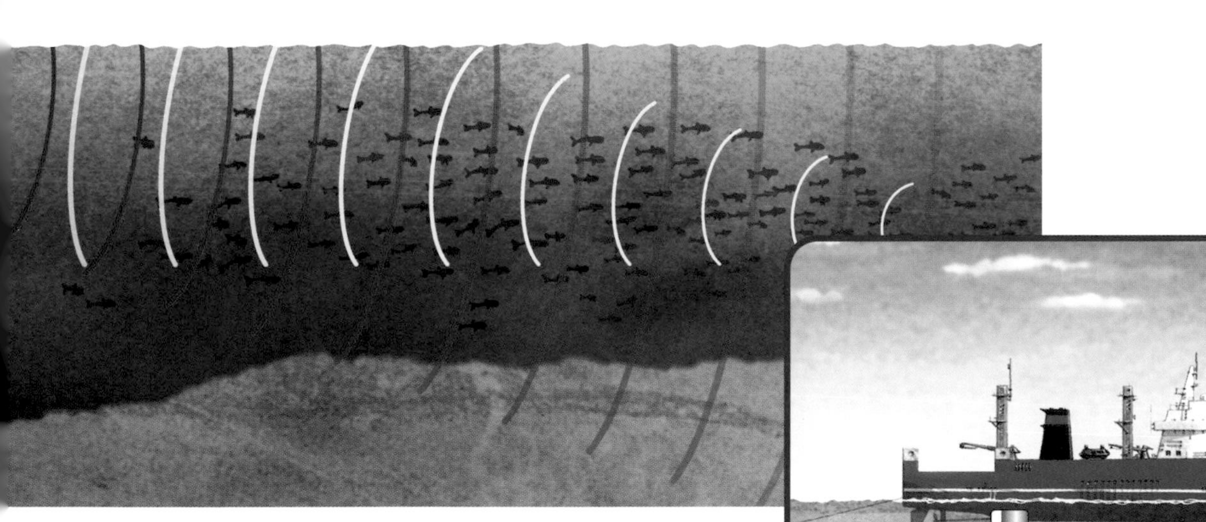

▲ **Echo sounding** uses the simplest type of sonar. An ultrasonic signal reflects off the seabed, and a detector on the sending vessel picks up the echo. The depth of water can be calculated from the time the signal takes to travel to the seabed and back.

Icebergs and submarines

A Frenchman, Paul Langevin (1872–1946), experimented with ultrasound at the beginning of the 20th century. His first idea for using it in echo sounding at sea was so that ships could avoid icebergs. But during World War I (1914–18) a new weapon appeared: the attack submarine. Torpedoes fired from submarines hiding beneath the waves sank many ships, and Langevin realized that the submarines could be located by using ultrasound. The technique was not perfected until after the war had ended, but it formed the basis

FOR MORE ON UNDERWATER ECHOES SEE *DETECTING AND RELAYING SOUND* **7**:*10*; *REPRODUCING SOUND* **7**:*20*; *ULTRASOUND–TOO HIGH TO HEAR* **7**:*24*

Underwater echoes

▲ **A nuclear submarine** prepares to dive and continue on its invisible patrol beneath the waves. Under the water it uses sonar to locate other submarines and surface vessels, as displayed on its sonar screens (inset).

for the sonar systems that soon came into use with the world's navies.

Sonar is short for **so**und **n**avigation **a**nd **r**anging. As well as submerged submarines it can detect surface vessels and underwater obstructions such as rocks and shipwrecks. It also picks up echoes reflected by whales and shoals of fish. The type used by fishing boats is sometimes called a "fish finder."

Sonar makes use of ultrasonic sound waves, which travel very well in water. Human beings cannot hear ultrasonic sounds because they are too high in pitch. With some sonar equipment the ultrasonic pulses are converted into audible tones that the operator can hear as a series of "pings." An experienced operator can guess the nature and size of an object from the sound of the pings it echoes back—a shoal of fish sounds very different from a submarine, for example. The sonar can also pick up the pings from other sonar transmissions. They reveal that another nearby vessel is making a sonar search.

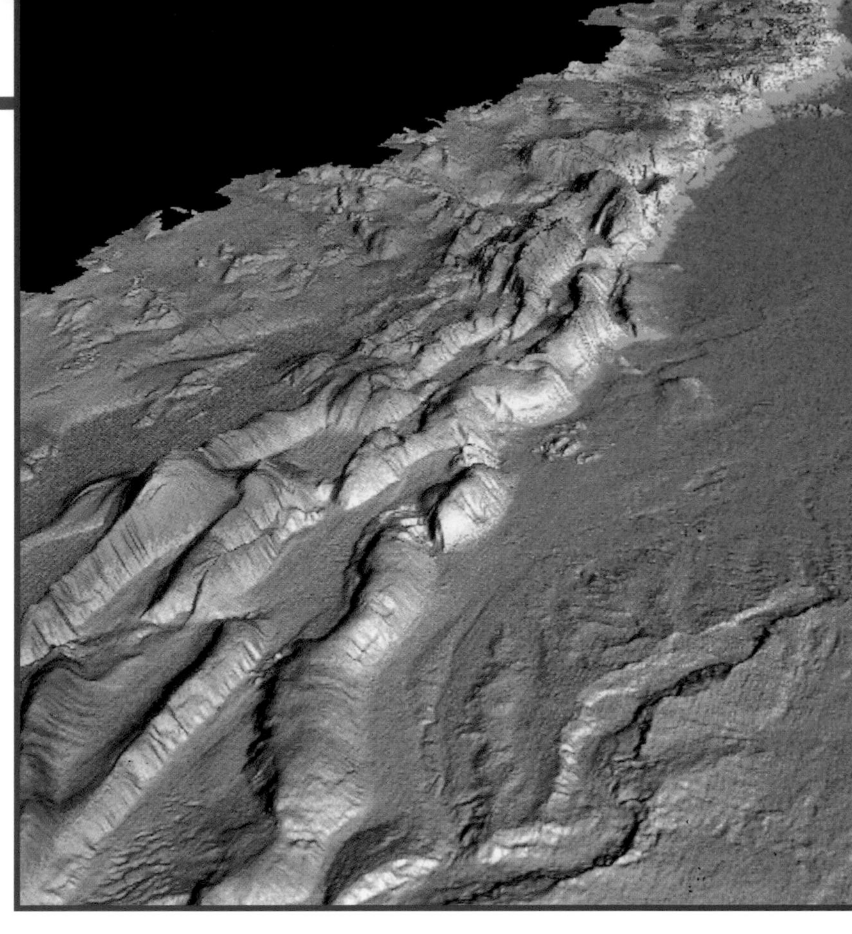

▲ **Mapping the seabed** is possible with high-definition sonar. This map, given "false" colors by a computer, shows the east Pacific seabed off the coast of Oregon (shown black; north is to the left).

Dunking sonars

The ship that sends the sonar signals can also itself reflect the returning echoes. To prevent the confusion that this can cause, and to increase range, warships can use another device. In what is called a dunking sonar the ship's helicopter carries a sonar transmitter and receiver at the end of a long wire. It hovers over the sea surface and lowers the wire into the water. Any returning signals pass back up the wire to the helicopter, which then transmits them by radio to the mother ship.

Passive sonars

The devices described so far are examples of active sonars. They actively send out supersonic signals. A passive sonar does not send out its own signals—it is merely a listening device. It can pick up the sounds of the engines and propellers of other ships and submarines. It also indicates the direction the noises are coming from.

Passive sonars have revealed that ships are not the only things that make noises underwater. Whales and dolphins make wailing cries with which they "talk" to each other. They also emit high-pitched clicks and squeaks, which they probably use in the same way as sonar to locate objects under the water. Active sonars on ships may interfere with this system.

▲ **The Atlantic ocean floor** in the Gulf of Mexico has also been mapped using sonar. This detailed map shows the continental shelf off the coast of Louisiana. It was created by a computer using data from surface ships that recorded echoes from the seabed.

Infrasound — too low to hear

The shock waves produced by an earthquake that travel through the ground and cause so much damage are in fact sound waves of very low frequency. We cannot hear them, but we can feel them when they make the ground vibrate as they pass and detect them with seismographs.

Human hearing has its lower limit at a frequency of about 20 hertz. Frequencies lower than that are called infrasound. The most dramatic form that infrasound can take is the seismic shock waves that are produced by earthquakes. If you need reminding that sound waves carry energy, just consider the destructive force of earthquakes!

Why earthquakes happen

During an earthquake the ground shakes because the Earth's crust is on the move. The crust—the Earth's outer layer—consists of about twenty huge plates that fit together rather like the tiles of a giant spherical mosaic. The plates "float" on the molten rock that forms the mantle, the next layer down. Convection currents in the mantle make the plates slowly move around and rub against each other at their edges. In some places two plates are colliding with each other, and one slides beneath the other. At other places the edges of two plates become locked together and then suddenly slide apart. Any of these plate movements can cause an earthquake.

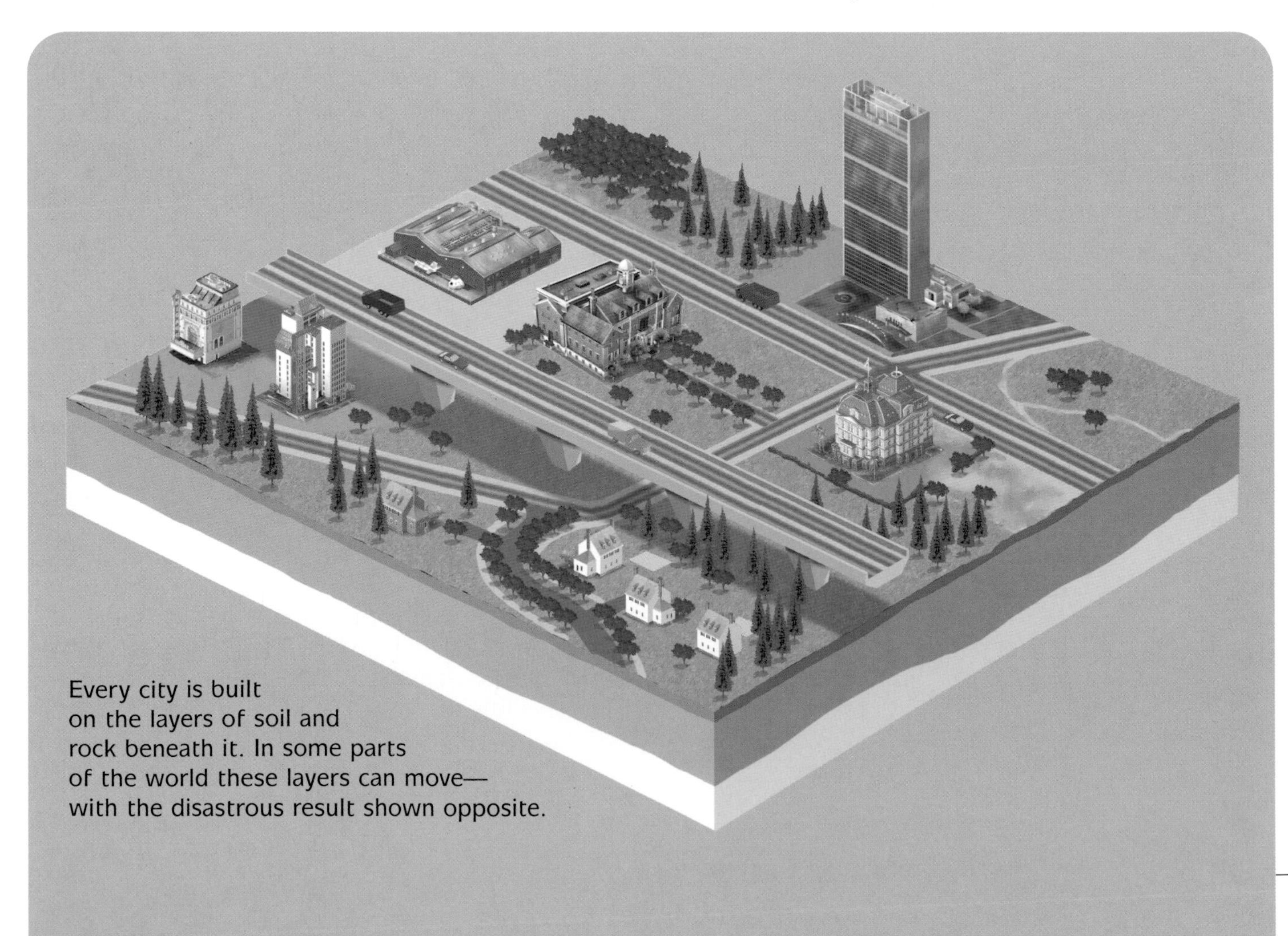

Every city is built on the layers of soil and rock beneath it. In some parts of the world these layers can move—with the disastrous result shown opposite.

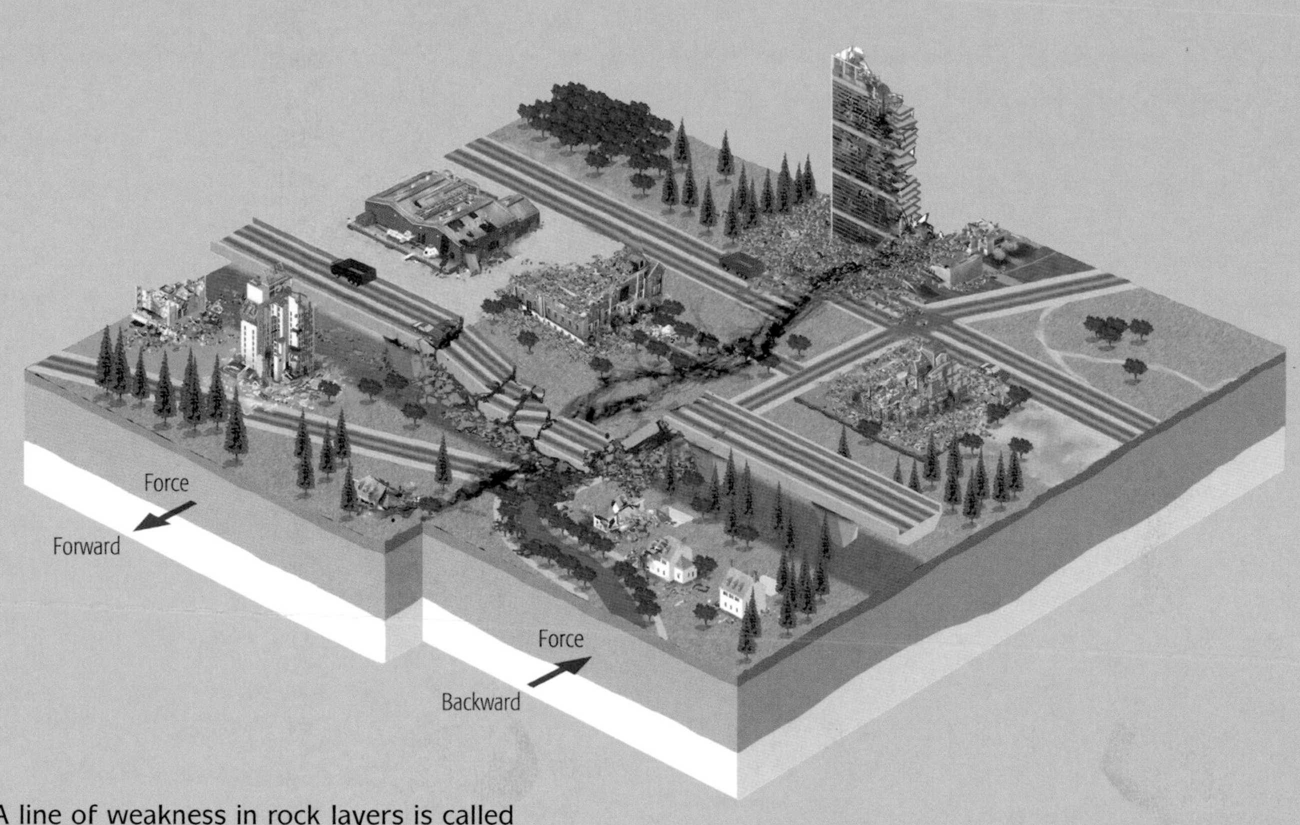

A line of weakness in rock layers is called a fault. Sideways movement along a fault results in an earthquake in which part of the land is displaced sideways.

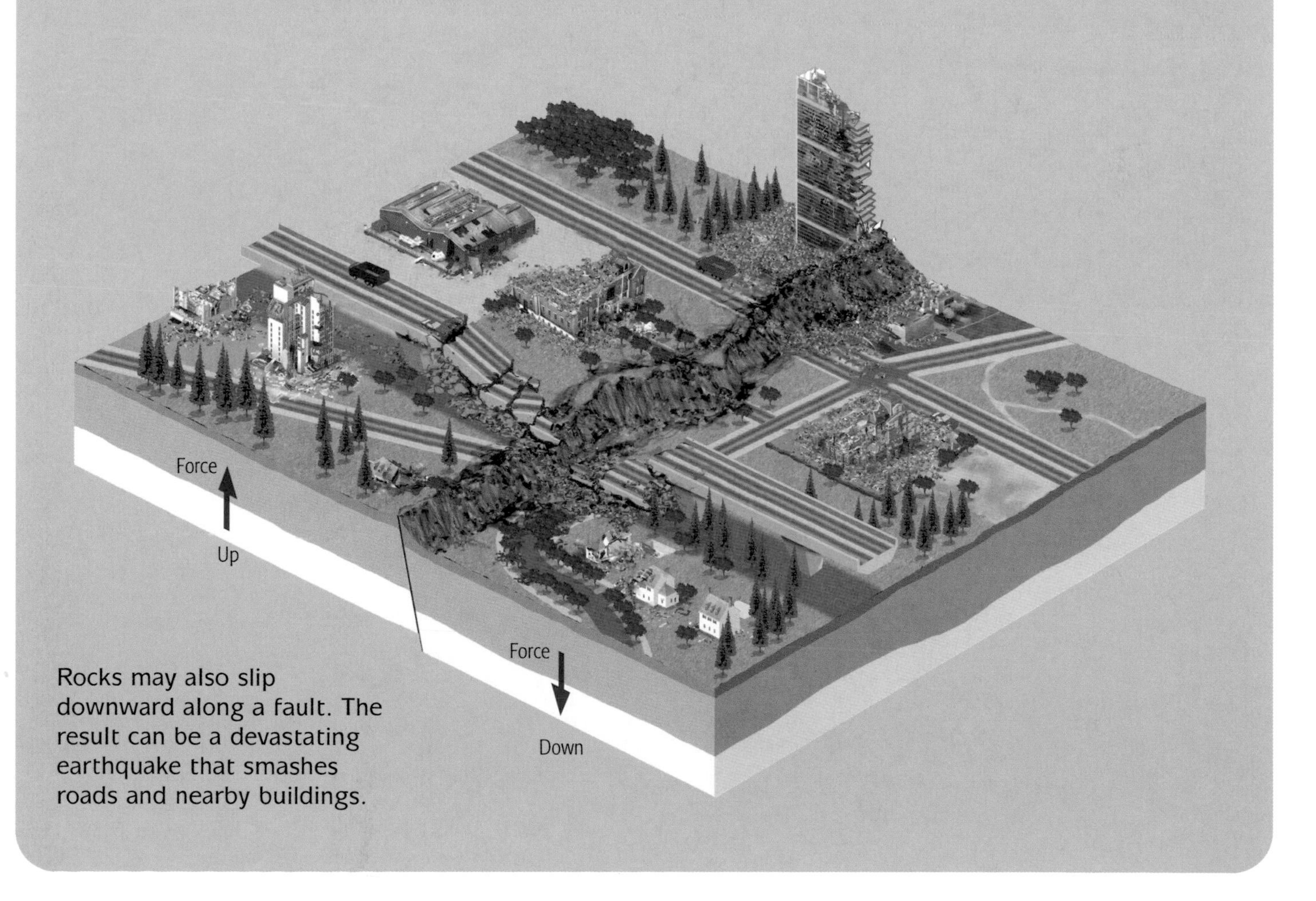

Rocks may also slip downward along a fault. The result can be a devastating earthquake that smashes roads and nearby buildings.

For more on infrasound see *Measuring magnetic fields* **4**:36; *Ultrasound–too high to hear* **7**:24; *Rocks and minerals* **9**:14

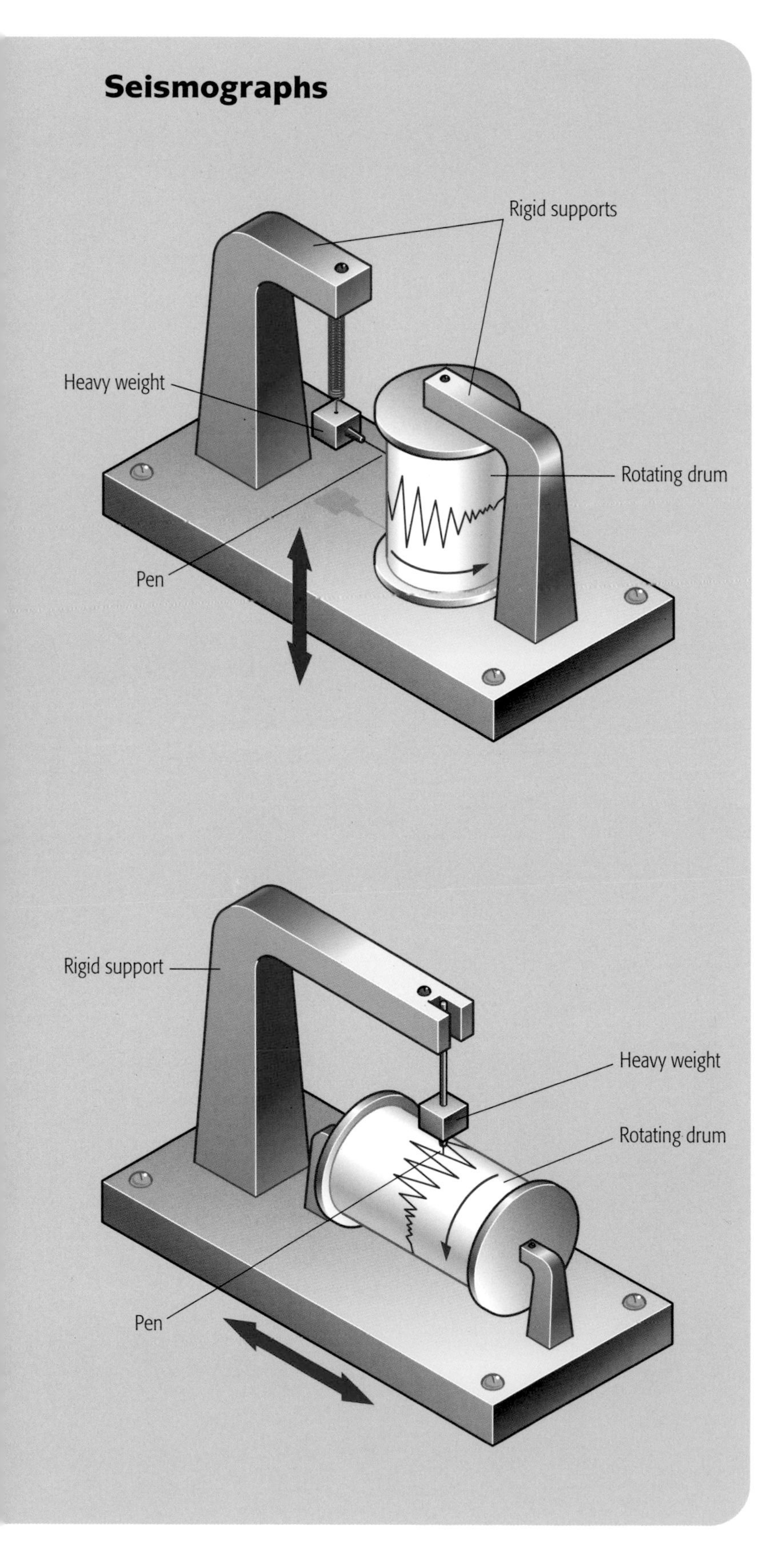

On a smaller scale more local movements can cause earthquakes. The layers of rock below the surface can include cracks called faults. Sometimes tension builds up along fault lines—perhaps because of nearby crustal plate movements. When this happens, the layers of rock may slip along the fault. They may slip sideways, or one may drop below the level of the other. The result is often an earthquake (see the illustrations on page 33). The most unstable fault lines are along the edges of two neighboring plates. An example is the San Andreas Fault in California, whose movement has caused several devastating earthquakes in the San Francisco area.

Detecting earthquakes

A scientific instrument for detecting the seismic waves from earthquakes is called a seismometer. If it also produces an ink trace on a roll of paper, it is a seismograph. Two common types are shown on this page. They both rely on inertia, which is the property of an object that makes it resist being moved. The more massive an object, the higher its inertia. Both seismographs consist of a heavy metal frame with supports for a pendulum-type weight and a rotating drum carrying a roll of paper. The frame is bolted to a large block of concrete with its base buried in the ground.

◄**A seismograph** for detecting up-and-down movements (top) has a weight suspended by a spring and carrying a pen. Because of its inertia the weight does not move when an earth tremor shakes the rest of the instrument up and down. The pen traces a record of the vibrations on a rotating drum of paper. A different type of seismograph (bottom) detects side-to-side movements. Again, the weight remains still while earth tremors make the rest of the instrument vibrate sideways.

◀ **The photographer** has shaken the camera to give an impression of what it feels like when the whole world seems to be in turmoil during an earthquake. The cause is infrasound waves traveling through the ground from the source of the earthquake.

The upper instrument is designed to measure up-and-down earthquake tremors. When the ground vibrates up and down, so does the frame of the seismograph. The inertia of the heavy weight at the end of its spring makes it tend to remain still, so the pen attached to it draws an up-and-down trace on the drum (which is attached to the frame and so moves with it).

The lower instrument measures side-to-side earthquake movements. In a similar way the inertia of its suspended weight tends to keep it still while its pen draws a trace on the rotating drum as the drum moves from side to side. Some seismographs, like the one illustrated on the right, have more than one pen. Other types are connected directly to a computer that records the information.

▲ **The pens on a seismograph** trace out wavy lines that indicate there has been an earthquake somewhere within recording range.

◀**An earthquake** is one of the most destructive forces in nature. Unless buildings are constructed to be "earthquake-proof," they can be literally shaken to pieces in only a minute or two of earth tremors. This earthquake in Mexico City registered 8.1 on the Richter scale.

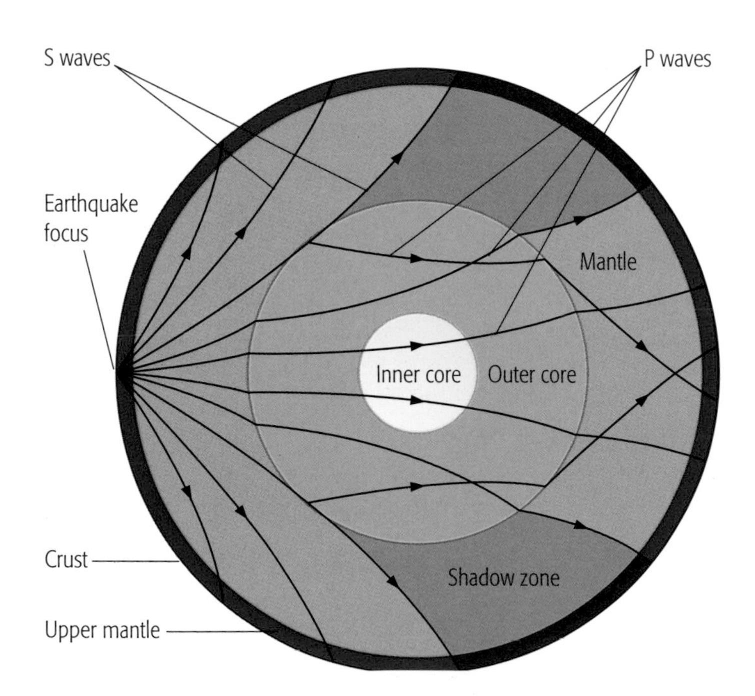

▲**Earthquake waves** travel at different speeds through different layers of the Earth. They bend because they change speed in passing from one layer to the next. These changes reveal the Earth's inner structure.

Earthquake waves

There are two main kinds of earthquake waves. Primary waves, known as P waves, travel as regions of compression (just like sound waves in air), causing rock particles to vibrate back and forth like a coiled spring in the direction of the wave. They easily pass through gas, liquid, and rock. P waves speed up as they pass through the Earth's mantle (see the illustration on the left) but slow down when they reach the outer core.

Shear waves, or S waves, on the other hand, make rock particles vibrate at right angles to the direction of the wave, like a vibrating guitar string. These waves will travel only through solids and for this reason do not pass through the Earth's molten outer core. As the waves pass through layers of rock of different density, they change speed. This has the effect of making their paths curve. As a result, there is a region called a shadow zone, about one-third of the way around the Earth in each direction from the focus of the earthquake, that no waves reach.

Charles Richter

Charles Francis Richter was an American seismologist who is best known for his scale for measuring the strength of earthquakes. He was born in Hamilton, Ohio, in 1900 and was educated at the University of Southern California, Stanford, and the California Institute of Technology, where he later became professor of seismology. He developed his earthquake scale in 1935. Unlike the earlier 1902 scale of Giuseppe Mercalli, which defines earthquakes in terms of the damage they do, Richter's scale is based on the magnitude of the earthquake measured in terms of the amplitude (strength) of the infrasound waves it produces in the ground. The strongest earthquake ever recorded (in Assam, India, in 1950) reached 8.9 on the Richter scale. Richter died in 1985.

Where both types of waves, P and S, reach the surface, they change into long waves, also called L waves. They travel up and down like waves on the sea or vibrate sideways along the surface of the ground. It is these waves that do the most damage in an earthquake.

Measuring earthquakes

The intensity of an earthquake is usually judged by the amount of visible damage it causes. This is the basis of the Mercalli scale, illustrated on the right. It grades earthquakes

Level	Description according to the Mercalli Scale	
I	Not felt; detected by seismographs	
II	Felt by a few people; hanging objects swing	
III	Felt by a few people; vibrations like a passing truck	
IV	Felt by many people; dishes and windows rattle noisily	
V	Felt by most people; sleepers are awakened, unstable objects may fall	
VI	Felt by everyone; walkers stagger, windows break, trees shake visibly	
VII	Hard to stand up; plaster and loose tiles fall, some chimneys break	
VIII	Car steering difficult, chimneys fall, partial collapse of buildings, branches break	
IX	People panic; serious builiding damage, foundations and underground pipes crack	
X	Most buildings destroyed, large landslides, water thrown out of canals and rivers	
XI	Railroad tracks bent, bridges and all underground pipework destroyed	
XII	Total destruction; large rocks moved, objects thrown in air	

Infrasound — too low to hear

◀ **Some spiders** are able to detect infrasound—sounds that are too low in pitch for humans to hear. More like vibrations in the ground, these infrasounds tell the spider when an enemy or a possible meal is approaching.

▶ **Setting off an explosive charge** in a borehole produces a small artificial earthquake. The behavior of the infrasound waves, as picked up by a series of seismographs, tells geologists about the structure of the rocks and the likely location of mineral deposits.

Chinese earthquake detector

In A.D. 132 the Chinese scientist Chang Heng invented an apparatus for detecting earthquakes. It was a large jar with six dragons' heads arranged around the sides. Each dragon held a ball in its mouth. Nobody knows the exact details of the mechanism inside, but it probably involved a pendulum. When there was an earth tremor, the pendulum swung to one side. That movement pushed a lever to open a dragon's mouth so that it let go of the ball, which fell into the open mouth of a waiting toad. The direction of the earthquake was indicated by which ball fell.

▶ **An early earthquake detector**, or seismograph, was constructed nearly 2,000 years ago in ancient China.

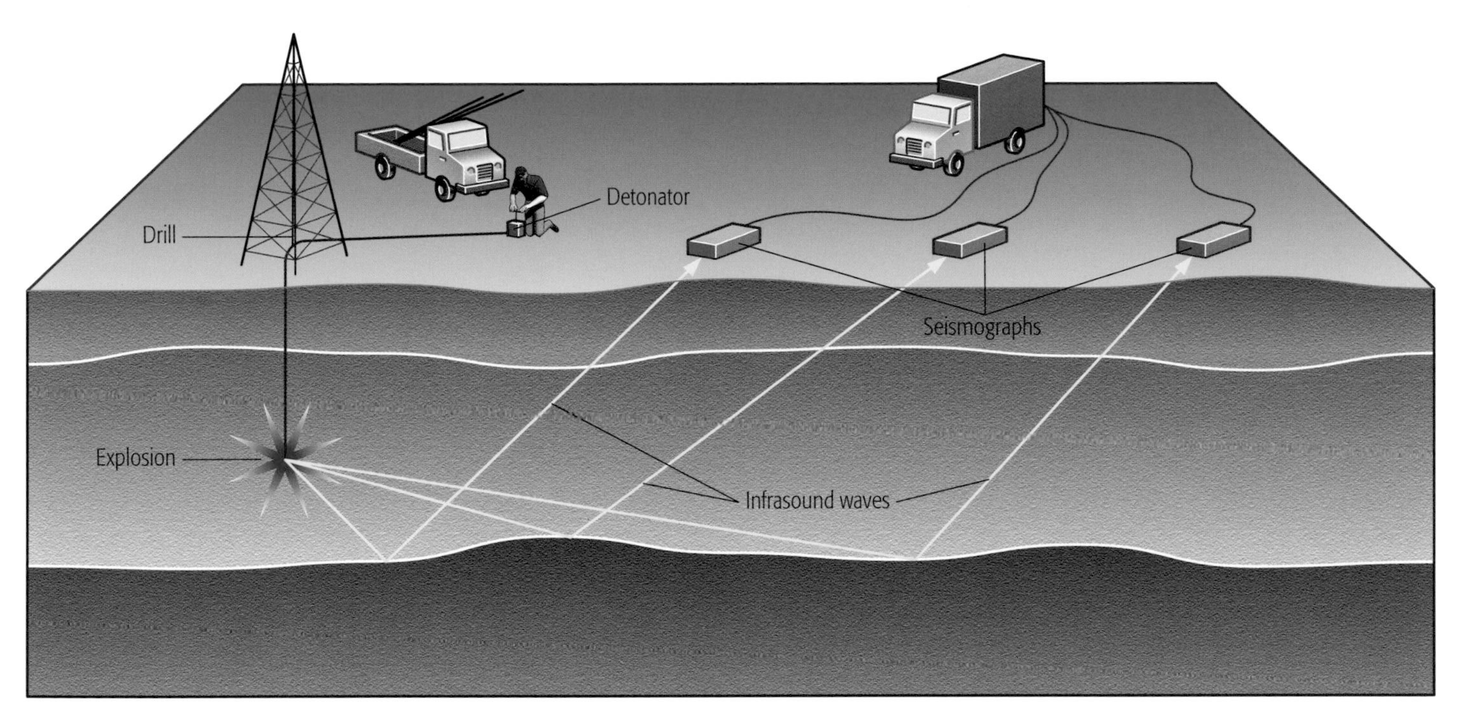

on a scale of I (earthquakes detectable only by seismographs) to XII (earthquakes that cause total destruction).

The alternate Richter scale measures the magnitude of an earthquake independently of what people experience or how much damage it causes. The energy is estimated from the amplitude of the trace on a seismograph, taking into account the distance to the focus of the earthquake (where it originates). The Richter scale runs from –3 for earthquakes involving very little energy to +9 for very destructive, high-energy earthquakes. It is a logarithmic scale, which means that an increase of 1 on the scale represents an increase by a factor of 10 in the energy of the earthquake. The worst earthquake ever recorded peaked at 8.9 on the Richter scale. The energy released was equivalent to 6 million tons of TNT.

Artificial earthquakes

Studying seismic waves from earthquakes can provide scientists with information on the structure of the Earth itself. For a more local study geologists make their own small earthquakes. They do this by boring a hole into the ground and setting off an explosive at the bottom of the borehole. The waves produced are picked up by a series of seismographs located on the surface (the setup is illustrated above). From the time the waves take to reach the instruments the geologists can work out details about the structure of the rocks they have traveled through. The method can also reveal the presence of minerals and is particularly important in prospecting for underground deposits of oil.

Animals and infrasound

Zoologists have proof that animals as diverse as elephants and spiders can detect waves of infrasound. This ability probably accounts for the strange behavior of some animals just *before* an earthquake occurs. Scientists in China—where there have been many destructive earthquakes—are studying this behavior, hoping to use it as a basis for predicting earthquakes and saving people's lives.

Glossary

Any of the words in SMALL CAPITAL LETTERS can be looked up in this Glossary.

acoustics The scientific study of sound and sound WAVES. The properties of a building such as a theater or concert hall that affect sounds (speech and music) produced within it are also called its acoustics.

amplifier An electronic device that increases the strength of an electric signal (e.g., the signal from a MICROPHONE).

amplitude The height (or depth) of a traveling WAVE measured from its mean position. It is a measure of the intensity of the wave.

analog signal An electric signal that varies continuously; an example is the signal from a MICROPHONE, which varies in step with the sound WAVES producing it.

bel (symbol **B**) A unit used to compare power levels such as the LOUDNESS of sound. In practice, the unit used is the DECIBEL (one-tenth of a bel).

binary number A number expressed using only two digits, 0 and 1. The functioning of digital computers and other digital devices such as COMPACT DISKS is based on binary numbers.

carbon microphone A type of MICROPHONE in which sound waves vibrate a DIAPHRAGM to vary the electric resistance of carbon granules. The varying resistance in turn varies the voltage from a battery, which is the output signal.

coaxial cable An electric cable consisting of a central conductor (wire) surrounded by an insulator and then by a tubular or braided second conductor.

compact disk (CD) An optical recording medium in which digitized sound signals are recorded as a series of microscopic pits on an aluminum disk, covered with a protective layer of plastic. Computer data and digitized images can also be recorded on compact disks.

condenser microphone A type of microphone in which sound waves vibrate a DIAPHRAGM to vary the capacitance of a condenser. The varying capacitance in turn varies the voltage from a battery, which is the output signal.

crystal microphone A type of microphone in which sound waves vibrate a DIAPHRAGM to vibrate a PIEZOELECTRIC CRYSTAL, which produces a varying voltage as the output signal.

decibel (symbol **dB**) One-tenth of a BEL.

diaphragm A thin sheet of metal or plastic that vibrates when struck by sound waves (as used in a MICROPHONE).

digital audio tape (DAT) A type of recording tape on which the sound signals are digitized, in the form of a DIGITAL SIGNAL as opposed to an ANALOG SIGNAL.

digital signal An electronic signal that consists of a stream of "on" and "off" pulses, corresponding to the binary digits 1 and 0, as on a COMPACT DISK.

echo A sound (or other wave) that is reflected from a surface, such as a cliff or large building. Echo sounders, SONAR and ULTRASOUND scanning all make use of echoes.

electrostatic speaker A type of LOUDSPEAKER in which varying electric charges on either side of a sheet of metal make it vibrate and produce sounds.

fault A line of weakness in a rock formation along which the rock may move (up or down, or sideways).

frequency For any WAVE motion, the number of complete cycles of the wave that pass a given point every second. It is reckoned in HERTZ.

hertz (symbol **Hz**) The SI unit of frequency. One hertz equals a frequency of one cycle of the WAVE or oscillation per second.

infrasound Sound that is too low in FREQUENCY (pitch) for humans to hear.

intensity The strength (loudness) of a sound wave, represented by its AMPLITUDE.

laser A device that produces an intense beam of monochromatic (single-color) light that is coherent (all of its waves are exactly in step with each other). "Laser" stands for **l**ight **a**mplification by **s**timulated **e**mission of **r**adiation.

loudness The way in which we perceive the INTENSITY of a sound.

loudspeaker A device for converting electric signals (for example from a MICROPHONE or tape recorder) into sounds.

Mach number The ratio of the speed of an object to the SPEED OF SOUND in the same medium.

magnetic tape Plastic tape that has a coating of metallic particles that can be magnetized to form a record of an electronic signal (analog or digital).

Mercalli scale A scale for measuring earthquakes in terms of the visible damage they cause.

microphone A device for converting sounds into electric signals (which can then be amplified and recorded).

moving-coil microphone A type of MICROPHONE in which sound waves vibrate a DIAPHRAGM to vary the position of a coil of wire in the field of a magnet. This movement of the coil generates a varying electric voltage in the coil, which is the output signal.

photoelectric cell An electronic device that generates an electric current when light falls on it.

piezoelectric crystal A type of crystal that generates an electric current when it is strained. A piezoelectric crystal of quartz is used in a CRYSTAL MICROPHONE.

pitch The highness or lowness of a sound (note), represented by its FREQUENCY.

primary wave A wave, produced by an earthquake, that takes the form of bursts of compression traveling in the direction of travel. Also called a *P wave.*

record head In a tape recorder an electromagnet that magnetizes metallic particles on the tape as it passes by to form a record of the input signal to the head.

repeater A type of AMPLIFIER inserted at regular intervals along a long stretch of telephone cable to boost the signal.

replay head In a tape recorder an electromagnet that produces an electric output signal when the magnetized (prerecorded) tape passes it.

ribbon microphone A type of MICROPHONE in which sound waves vibrate a DIAPHRAGM to vary the position of a length of metal foil (a metallic ribbon) in the field of a magnet. This movement of the foil generates a varying electric voltage in the foil, which is the output signal.

Richter scale A scale for measuring earthquakes in terms of their energy output (as recorded by a SEISMOGRAPH).

seismograph An instrument for detecting and producing a chart record of earthquake waves.

shear wave A wave produced by an earthquake that travels in the form of a series of sideways vibrations along the direction of travel. Also called an *S wave.*

sonar An electronic device that detects the range and bearing (direction) of underwater objects, such as submarines, which reflect back echoes of the ULTRASOUND signals that the sonar transmits.

speed of sound The speed of sound in dry air, equal to 344 meters per second (about 1,130 feet per second).

synthesizer An electronic instrument that can reproduce the sound of any musical instrument (and many other sounds besides).

threshold of hearing The quietest sound we can hear, assigned a value of 0 DECIBELS.

transducer A device that converts one kind of energy into another kind, particularly a nonelectrical signal into an electrical signal or vice versa.

tweeter A type of LOUDSPEAKER that is designed to accurately reproduce high-pitched sounds.

ultrasound Sound that is too high in FREQUENCY (pitch) for humans to hear.

ultrasound scanner A machine for investigating body tissues or other structures by recording the echoes of ULTRASOUND waves reflected by them.

vibration A rapid back-and-forth movement of a material or object. Anything that produces sounds must vibrate.

wave A regular disturbance in space or within a medium that transfers energy in its direction of travel. For example, sound travels as waves.

wavelength The distance between two neighboring peaks (or troughs) of a traveling WAVE.

woofer A type of LOUDSPEAKER that is designed to accurately reproduce low-pitched sounds.

Set Index

Numbers in **bold** in this index refer to the volume number. They are followed by the relevant page numbers. A page number in *italics* indicates an illustration.

A

accumulator **4**:21, *22*, 32
acid strength **8**:17, *18*
actinides **9**:*12*
air **3**:8, 22
air traffic controllers **7**:11
airplanes **3**:22, 25; **6**:18, **7**:9
airports **6**:27
Alfred the Great **2**:12
algae **5**:*19*, *20*, **9**:23, *32*
alpha particles **3**:36, *37*
altimeter **1**:23, *23*, **3**:25
altitude **1**:23
aluminum **9**:12
aluminum oxide **9**:14
aluminum silicate **9**:14
ammeter **4**:14
  hot-wire **4**:16, *16*
  magnets **4**:15, *15*
  moving-coil **4**:15, 15
  moving-iron **4**:16–17, *17*
  ohmmeter **4**:29, *29*
  resistance **4**:15, 20, *20*
ammonium carbonate **8**:*12*, 14
ammonium hydroxide **8**:*12*, 13, 19, *19*
ampere **4**:7, 9, 14
Ampère, André-Marie **4**:6, 7, *19*
Ampère's law **4**:*19*
amphibians **9**:25, *25*
amplifier **7**:16
amplitude **5**:7, 12; **7**:7, *7*, 17–18
analog signals **7**:17–18, *18*, 19
analysis **8**:*6*, 6–9
  gravimetric **8**:8, 16, 18–19, *19*
  qualitative **8**:*6*–9, *7*, *12*
  quantitative **8**:6–9, *8*, *16*, 16–19, *17*, *19*
  volumetric **8**:8, *16*, 16–18, *17*
  wet **8**:*7*, *12*, 13, 24
Andromeda **9**:*34*
anemia **9**:30
angiogram **6**:29
angiosperms **9**:20–21
angles **1**:20, *20*, 21, 22
  heights **1**:20–23
  parallax **1**:*18*, 18–19
  tangents **1**:21
  theodolites **1**:14
angstroms **5**:7
animals **9**:24, 25–29, *27*, *32*
  infrasound **7**:*38*, 39
  ultrasonic sound **7**:26, *27*
anode **4**:13; **8**:15
ants **8**:21
Apollo 12 **4**:*33*
arachnids **9**:8, 25, 29
area **1**:24–27, *25*, *26*, *27*
Arecibo telescope **6**:*39*
argon-40 **2**:36; **8**:15
Arizona **2**:*35*; **5**:32
arthropods **9**:25, 28–29
asteroids **9**:38
Aston, Francis **8**:26
astrology **2**:11
astronomy
  Babylonians **2**:9
  classification **9**:34–39
  clocks **2**:14–15, 21
  Egypt **2**:10
  infrared **6**:14, 18–19
  Maskelyne **2**:28
  micrometers **1**:12
  spectroscopes **8**:7, 22–23
  ultraviolet **6**:13
  units **1**:13, 14, 18, 19
  zodiac **2**:11
  see also radio astronomy; telescopes
atmosphere *see* Earth, atmosphere
atmospheric pressure **1**:23; **3**:8–9, 23–24
  isobars **1**:37
  lift pump **3**:25
atom
  acceptors/donors **3**:32–33
  bonds **3**:32–33
  crystals **3**:33, *33*
  electrons **6**:12
  heat **8**:23
  isotopes **3**:38–39, *39*
  nucleus **2**:34, **3**:36, *36*, **6**:12; **8**:23
  relative atomic mass **9**:10, 11
atomic mass units **3**:7
atomic number **9**:13
atomic spectra **8**:22–23
averages, law of **8**:33

B

baby **3**:18, 19; **7**:*24*
Babylonia **2**:9
bacillus **6**:23; **9**:33, *33*
bacteria **6**:*23*, *25*
  classification **9**:*32*, 32–33, *33*
  diseases **9**:32
  microscopes **5**:16, 21
  ultrasound **7**:26
Bain, Alexander **2**:21
balances
  bathroom scales **3**:19
  beam **3**:*16*, 17
  chemical **3**:*16*, 17
  Greek **3**:18
  kitchen scales **3**:16, *16*
  laboratory **3**:*17*, 17–18
  microbalance **3**:*17*, 17–18
  spring **3**:*18*, 18–19, *19*
  torsion **3**:11, *11*
balloon **3**:20; **6**:18
bar, unit **3**:9
bar chart **8**:*33*, 35
barium **8**:14, 21; **9**:11, *12*
barograph **3**:9, *23*, 25
barometer
  aneroid **1**:23, *23*; **3**:25
  banjo **3**:24, *25*
  Fortin **3**:24
  mercury **3**:9, 23, *24*
  Torricelli **3**:23–24, *24*
bats **6**:33–34; **7**:24, 26, *27*; **9**:26
battery
  clocks **2**:21, 25
  electric circuit **4**:*6*–*7*
  electrolysis **8**:15
  meter bridge **4**:26, *26*
  Volta **4**:8, *23*
  voltage **4**:21, *22*
beams
  electrons **6**:26
  infrared radiation **6**:15
  light **5**:8, 10, 11
Beatles **7**:*13*
bee-eaters **9**:*28*
bees **6**:7
beetles **9**:*7*, 8, *9*
Bell, Alexander Graham **7**:*9*, 14
benzene **7**:*22*; **8**:25
berries **4**:20, *20*
beryllium **3**:36; **6**:19; **9**:14
Berzelius, Jöns **9**:10
binary numbers **7**:*18*
Binnig, Gerd **6**:25
binoculars **5**:*26*, 26–27
binomial nomenclature **9**:23, *23*
biology
  cameras **5**:34
  classification **9**:7–8, 9
  microscopes **5**:16, 21
  radio receivers **8**:*32*
birders **5**:*26*, 27
birds **9**:25, *25*
bird's-eye view **1**:*32*–*33*
bivalves **9**:28
black-eyed Susan **9**:*21*, 23
black hole **6**:18; **9**:36, 38
Black Rock Desert **3**:*22*
blood **8**:29; **9**:30, 31
blood groups **9**:*30*, 30–31
blood pressure **3**:30, *31*
blood transfusions **9**:30, 30, *31*
Boisbaudran, Paul-Emile de **9**:12
bolometer **6**:15–16
borax bead test **8**:21, *21*
boron **9**:12
botulism **6**:*23*; **9**:32
Bourdon, Eugène **3**:21
Bragg, William Henry **3**:35
Bragg, William Lawrence **3**:35
brain disorders **7**:25
brain waves **4**:*24*, 24–25
braking systems **3**:30
bromine **9**:11, 12
bromoethane **8**:26, *26*
Bryophyta **9**:22
bubble chamber **3**:*37*
bugs **1**:*7*, 10; **9**:7, 29
bulbs, electric **4**:6, 7, 30, 34, *35*; **5**:7
bullet **2**:*30*, 31
Bunsen, Robert Wilhelm **8**:*22*, 23
Bunsen burner **8**:20, *22*, 23
burette **8**:*16*, 17
butcher's steelyard **3**:18
butterflies **5**:11, *13*; **6**:7; **9**:*6*, *9*, 29

C

cable television **6**:16
cables **4**:7; **7**:15
Cady, Walter **2**:25
Caesar, Julius **2**:10
calcium **6**:7; **8**:14, 21; **9**:11
calcium salts **9**:17
calculators **4**:*14*
calendars **2**:10–11
California **3**:*27*; **5**:*37*; **7**:34
calipers **1**:*8*, 9, *9*, 10
call center workers **7**:*10*, 11
camcorders **6**:16
cameras
  nondigital **5**:34–35
  single-lens reflex **5**:35–36, *36*, *37*
  space program **5**:*34*
  ultrahigh-speed **2**:31
  ultraviolet **6**:13
  underwater **5**:34
  see also photography
Canaveral, Cape **2**:30; **3**:14–15
Candida fungus **6**:*20*
capacitance **4**:8; **7**:*22*
capacitors **4**:8
capsule **9**:*20*
car **2**:32–33; **8**:37
carbon **2**:36; **3**:38; **9**:14
carbon-14 **2**:36–37; **3**:38
carbon dioxide **2**:36–37; **9**:18
carbonates **4**:13; **8**:*12*, 14, 19
carbonyl group **8**:25
carnivores **9**:26, *26*, *27*
Cassegrain, Laurent **5**:29
CAT scan **6**:28, 30, *30*, *31*; **7**:25
cathode **4**:13; **8**:15
cathode-ray tube **4**:22, *24*
cathode rays **3**:36; **6**:26
cats **1**:*10*; **9**:*24*, 26, *27*
Cavendish, Henry **3**:*11*, 11–12
CD player **7**:19, *19*
CDs **5**:11, *12*; **7**:16, 18, *19*
cellphone network **6**:34–35, *34*–*35*
cells **5**:16
census **8**:34
centimeters **1**:6, 7, 13, 24, 29
centrifuge **8**:*17*
cerium **9**:10
cesium **8**:23
CFCs **6**:10
Chadwick, James **3**:36
chain **1**:14
Chang Heng **7**:38
charge-coupled device **6**:16
Charles V, France **2**:18
Charles V, Spain **2**:21
chart **8**:34
cheetah **9**:24
Chelonia **9**:25
chemicals **9**:15
chemistry
  analysis **8**:6–9
  computer modeling **8**:37
  laboratory **9**:*11*
  x-rays **6**:27
  see also Periodic Table
Chicago University **5**:24
China
  clocks **2**:13, 14–15
  earthquakes **7**:39
  seismograph **7**:*38*
  sundials **2**:17
chlorides **3**:33, *33*; **8**:13, 19
chlorine atom **8**:27; **9**:11
chlorophyll **6**:*14*; **8**:*15*
cholera **9**:33
chordates **9**:25, *25*, *27*
chromatic aberration **5**:25
chromatography **8**:7, 10–15, *13*
Chromis viridis **9**:*29*
chronograph **2**:29
chronometer **2**:*26*, 26–27, *30*
Chrysanthemum leucanthum **9**:*21*, 23
circles **1**:26, *26*
circuit diagrams **4**:6–7
clapper board **7**:*16*
classes **9**:25
classification **9**:6–*7*, 6–9
  animals **9**:24–29
  astronomy **9**:34–39
  bacteria **9**:*32*, 32–33, *33*
  biology **9**:7–8, 9
  elements **9**:10–13
  fingerprints **9**:*8*, 9
  minerals **9**:*15*, *16*, 16–17
  plants **9**:18–23
  rocks **9**:17
clocks
  ammonia **2**:23
  aromatic **2**:13
  astronomical **2**:21
  atomic **2**:24, 39
  battery-driven **2**:21

candles **2**:12–13
cesium atomic **2**:*22*, 23
electric **2**:21
electronic **2**:*22*
gear wheels **2**:20–21
hourglasses **2**:12, *13*
hydrogen laser **2**:*23*, 24
hydrogen maser **2**:24
mechanical **2**:18–21
pendulum **2**:18–19
quartz **2**:*24*, 24–25, *25*
sandglass **2**:12
spring-driven **2**:20–21
24-hour **2**:*9*
water clocks **2**:12, 13–15, *14*
clockwork spring **2**:*19*
coccus **9**:33, *33*
colliders **3**:37–38
collimator **2**:23; **5**:10, *10*
colors
light **5**:8, *8–9*, 9
rainbow **5**:*6*, 9, *9*
reactants **8**:18
ultraviolet light **6**:*6*, 7
wavelengths **5**:*6*, 7
Colstrip, Montana **4**:9
Columbus, Christopher **2**:12
comets **9**:38
compass **1**:*35*
back bearings **1**:39
bearings **1**:20, 39
magnetic field **4**:36
maps **1**:*38*, *39*
complex shapes **1**:27, *27*, 30, *30*
Compositae family **9**:*21*, 23
computer modeling **8**:36–39, *37*
computers **8**:8–9
distances **1**:14
DNA sequences **8**:*31*
electroencephalogram **4**:*25*
LEDs **6**:16
PET scan **6**:*31*
concentrations **8**:9, *16*
condenser
electric charge **4**:8
microphones **7**:12, *12*
microscopes **5**:*17*, 19
condenser, distillation **8**:*10*, 12
conductivity, thermal **3**:21
conductors **4**:6, 7, *10*, 14, 26
cone-bearing plants **9**:21, 23
cones **1**:26, *26*, *29*
Coniferophyta **9**:22
constellations **2**:11; **9**:34–35
contours **1**:37
conversion tables **1**:6, 16; **3**:7
Coolidge, William **6**:26
coordinates **1**:35
Copernicus, Nicholas **5**:*24*
copper **4**:6, 7; **8**:21, **9**:10–11
copper ions **4**:13
copper sulfate **4**:13
corundum **9**:14, 16
cotyledon **9**:21
Coulomb, Charles de **4**:6, 7
coulombs **4**:7
coulometer **4**:13
covalent bonds **3**:32, *32*
cows **9**:26
Cray 2 supercomputer **8**:*38*
Crocodilia **9**:25
cross-sections **1**:30
crustaceans **9**:25, 29
crystallization **8**:12
crystallography **3**:*34*; **9**:16
crystals **2**:25; **3**:33, *33*, *34*, 35
Ctesibius **2**:14; **3**:26
cube **1**:29, *29*
cuboid **1**:29, *29*
Curie, Pierre **2**:25; **7**:12
customary units **1**:6; **3**:6, 8, 9
Cycadophyta **9**:22
cycads **9**:*22*
cyclotron **3**:37
cylinders **1**:26, *26*, 29, *29*

D

Dalton, John **9**:*10*
dandelion **9**:*21*
data presentation **8**:33, *33*, *34*, 35
day **2**:6, 8, *8*, 9, *9*
decibels **7**:6, 7, *8*
deflection angle **4**:18, *18*
demographic maps **1**:37
density **3**:8
depth **1**:6–7, *7*; **3**:9, 13; **7**:28
Dewey, Melvil **9**:8
Dewey decimal system **9**:8–9
diamond **3**:32; **6**:24; **9**:14, 16, 17
diaphragm
cameras **5**:34–35, *35*
microphones **7**:10, 11, 12, 14
pressure gauge **3**:21
diatoms **5**:*19*
dice rolling **8**:*34*, 35
dicotyledon **9**:21, *22*, 23
diffraction
light waves **3**:35, *35*; **5**:10, 11, *11*
sound waves **5**:12
diffraction grating **5**:11, 12, *13*; **8**:22
Digges, Leonard **5**:22
digital meters **4**:*34*
digital recording **7**:*18*, 19
digital signals **7**:17–18, 19
digital timers **2**:25, 29–30, *31*
dinosaurs **2**:34, *36*
diplococcus **9**:33, *33*
disco **7**:*20*
diseases **6**:7, 10; **9**:32, 33
displacement **1**:30, *31*
dissolving **3**:30
distance **1**:6–7, 14, 18, 32
distillation **8**:*10*, 12
dividers **1**:*9*
dividing engine **2**:21
DNA **3**:35; **5**:*15*; **8**:*28*, 28–31, *31*
Döbereiner, Johann **9**:11
dolphins **7**:31; **9**:26
dosimeter **4**:12
drams **3**:6
drupe **9**:20, *20*
dunking sonars **7**:31

E

earphone **7**:22, *22*
ears **7**:10, 32
Earth **9**:38, *38–39*
atmosphere **3**:9; **6**:10, *11*, 14, 18, 36
crust **2**:34; **7**:32
escape velocity **3**:*14*, 15
gravity **3**:15, 18
magma **9**:17
magnetic field **4**:36
mantle **7**:32, 36
measurements **1**:*19*; **3**:11–12
orbit **2**:6; **3**:12; **9**:36
poles **4**:36
pollution **5**:33
rotation **2**:6, 8, 23, 24, 39; **5**:25
time **2**:34–37
earthquakes
artificial **7**:*39*
detection **7**:*34*, 34–35, *35*
faults **7**:*32*, *33*, 34
India **7**:*37*
Mercalli scale **7**:37, *37*, 39
Mexico City **7**:*36*
photography **7**:*35*, *36*
plate movement **7**:32, *32*, *33*
Richter scale **7**:39
sound waves **7**:32
waves **7**:*36*, 36–37
echoes **6**:33–34
underwater **7**:28–31
eclipse **5**:*37*
Edison, Thomas Alva **4**:*35*
egg timers **2**:12
Egypt
astronomy **2**:10
calendars **2**:10
clocks **2**:13, *14*, 15–16
hours **2**:9
lift pump **3**:29
Nile River floods **3**:13–14
pyramids **1**:*31*
Thoth **2**:13
electric charge **4**:10–13
electric circuit **4**:6, *6–7*
electric current **4**:6
charged ions **4**:13; **8**:15, 26–27
heating effect **4**:14, 16, 34–35
magnetic field **4**:15, 16–17, 18, *18*
electric meters **4**:32–33, *34*
electric motor **6**:*27*
electric power plant **4**:*21*
electrical appliances **4**:*33*
electrical energy **3**:32–33
electricity **4**:32–35
conductors **3**:21; **4**:6, 7, 14
crystal **2**:25
resistance **4**:7, *27*; **6**:16
units **4**:6–9, 32
electricity substation **4**:*8*
electrocardiograph **4**:23–24, *24*
electrodes **6**:12, *12*
electroencephalograph **4**:25, *25*
electrolysis **3**:32; **4**:13, *18*; **7**:*22*; **8**:15
electromagnetic radiation **6**:9
spectrum **6**:6–9, *8–9*, 18, 26
wavelength **6**:6–9, 19
see also infrared radiation; ultraviolet radiation
electromagnetism **4**:*19*, 37; **7**:*14*
electrometer **4**:*11*, 12, *12*
electron gun **4**:22, *24*
electron micrograph **9**:*33*
electron probe microanalyzer **6**:24
electronic timers **2**:30–31
electrons
beams **6**:26
conductors **4**:14
copper **4**:6
electric charge **4**:10
excited **6**:12; **8**:23–24
ions **8**:26–27
orbits **3**:36, *36*; **8**:23
phosphorescent screen **6**:21–22
electrophoresis **8**:*28*, 29–30, *30*
electroplating **4**:13, *18*, *19*
electroscope **4**:10–12, *11*
elements
classification system **9**:10–13
Dalton **9**:*10*
discovery **8**:22–23
symbols **9**:10–11
elephant **3**:*9*
emerald **9**:14
emulsions **7**:26
engineering **1**:11, *13*; **8**:36
engineer's chain **1**:14
epilepsy **4**:25
equinox **2**:*7*, 8, 11
Eratosthenes **1**:*19*
escape velocity **3**:*14*, 15
escapements **2**:18–20, *18–20*
Escherichia coli **6**:25; **9**:*33*
European Space Agency **6**:19
evaporation **8**:*10*, 12
Extreme Ultraviolet Explorer **6**:13
eyeglasses **5**:14
eyes **5**:14; **6**:*24*

F

families **9**:26
Faraday, Michael **4**:8; **7**:*22*
farads **4**:8; **7**:*22*
faults, geological **7**:*32*, *33*, 34
feet **1**:6–7, 13, 16
Felidae family **9**:*24*, 26, *27*
femtoseconds **2**:31
Fermat, Pierre de **8**:*35*
ferns **9**:22–23
fiber-optic cables **6**:16; **7**:15
filaments **4**:7
Filicophyta **9**:22
filtration **8**:10, *10*
fingerprints **8**:8, *9*; **9**:*8*, 9
genetic **8**:*28*, 28–31
fireworks **8**:21
fish **9**:25, *25*, 26, *29*
fishing **7**:*28–29*, 30–31
flagella **9**:33
Flagstaff, Arizona **5**:*31*
flame tests **8**:20, 21
fleas **3**:*8*
floods **3**:13–14
fluorescent screens **6**:*16*, 16, 26, 28
fluorescent tubes **6**:10, *12*, 12–13
fluorite **6**:13; **9**:16
focal length **5**:*14*, 23–24
focus **5**:14–15, 20, 30
food poisoning **6**:*23*
forensic science **8**:8, *9*, 29
Forest, Lee de **7**:19
Fortin, Jean **3**:24
fossils **2**:34, *36*
fractional distillation **8**:*10*, 12
fractionating column **8**:*10*
France **2**:11; **3**:39
Frankenstein film **4**:*17*
frequency **5**:7
electromagnetic radiation **6**:9
gamma rays **6**:9
light waves **5**:8
radio waves **6**:9
sound waves **7**:6–9, *8*
Fritz Haber Institute **6**:25
fruit **9**:19, 20, *20*
Fundy, Bay of **3**:13

Set Index

Fungi **9**:23, *32*
furlong **1**:14, 16
fuses **4**:30

G

galaxies **3**:7; **6**:38; **9**:34, *34*, 36, *36*, 37, *37*
Galileo Galilei **2**:19, 28, **3**:*24*, **5**:22, *24*
see also telescopes
gallium **9**:12
gallium arsenide **6**:16
galvanometer **4**:18, 21, *22*, 26, *26*
tangent **4**:18, *18*, 37
gamma rays
Earth's atmosphere **6**:10, *11*
frequency **6**:9
production **6**:27, 30
radioactivity **6**:6–7, 27
Sun **6**:10, *11*
wavelength **6**:6–7
gas chromatography **8**:*14–15*
Gascoigne, William **1**:12
gases
conductivity **3**:21
density **3**:8
pressure **3**:20–25
gastropods **9**:28
gauges **1**:*9*, 11, 12, *12*
strain gauge **2**:25
see also pressure gauges
gear wheels **2**:20–21, *21*
gears, car **2**:32–33
Gell-Mann, Murray **3**:36
genera **9**:26
General Electric Company **4**:*35*
generators **3**:14; **4**:9
genetic fingerprinting **8**:28, 28–31
genetics **8**:8
genus **9**:18
geology
faults **7**:*32*, *33*, 34
Grand Canyon **2**:*34*
microscopes **5**:21
see also earthquakes; minerals; rocks
geometry **1**:21
germanium **9**:12
geyser **3**:*22*
gigawatt **4**:9
ginkgo **9**:22
Glaser, Donald **3**:37
Global Positioning System **1**:39
gluons **3**:*36*, 37
Gnetophyta **9**:22–23
gold **3**:8
Graham, George **1**:12
Gram, Hans **9**:33
grams **3**:6
Gram's stain **9**:33
Grand Canyon **2**:*34*
granite **2**:*34*; **9**:17
Grant, George **2**:21
gravity **3**:10
Earth **3**:15, 18
effects **3**:12–13
mass **3**:6, 11
Moon **3**:15, 18
Newton's law **3**:*10*, 11, *13*
rocket launch **3**:*14*
Solar System **3**:15
weight **3**:6–7
Greeks
balance **3**:18
Ctesibius **3**:26
sundial **2**:16–17
water clock **2**:14, *14*
Green, Andy **2**:32
Greenwich Mean Time **1**:35, **2**:26, *30*, *39*
Gregory, James **5**:29
Gregory XIII, pope **2**:10
Grimaldi, Francesco **5**:11
guitar **7**:23
Gunter's chain **1**:14
Guo Shoujing **2**:17
gymnosperms **9**:21–22

H

hair, static electricity **4**:10, *10*
hairsprings **2**:20; **4**:15, *15*, 17, *17*
Hale, George Ellery **5**:29
halite **3**:*33*
hardness test **9**:16–17
Harrison, John **2**:28, *29*
Hawaii, Mauna Kea **5**:*30*, 32; **6**:19
headset **7**:*10*, *22*
hearing aids **7**:11
heart **3**:30; **4**:23
heartbeats **4**:23–24
heat
infrared radiation **6**:8
heat rays **6**:8, 10, 14
heating effect **4**:14, 16, 34–35; **8**:23
heights **1**:6–7, *7*, 20–23, 25
helium **3**:36; **8**:*14*, 15
Henlein, Peter **2**:20
Henry, Joseph **4**:8
Henry, Sir Edward **9**:9
henrys **4**:8
herbivores **9**:26
Herschel, Caroline **6**:*15*
Herschel, Sir William **5**:29; **6**:15, *15*
hertz **5**:8; **6**:9; **7**:6, 8, 21, 24
Hertz, Heinrich **7**:8
Hertzsprung, Ejnar **9**:35
Hertzsprung-Russell diagram **9**:35, *35*, 36
hesperidium **9**:20, *20*
hikers **1**:*35*
Hillier, James **6**:25
Hipp, Matthäus **2**:21
Hooke, Robert **2**:20; **3**:24; **5**:16
horizon glass **1**:22
horoscopes **2**:11
horse **9**:*27*
horse races **1**:16
horsetails **9**:22
Horton, Joseph W. **2**:24
hourglass **2**:12, *12*, *13*
hours **2**:9
Hubble, Edwin **9**:*36*, 37, *39*
Hughes, David Edward **7**:14, *14*
humans
blood **9**:30–31
evolution **9**:9
hearing **7**:32
teeth **9**:*27*, 28
hundredweights **3**:6
hurricanes **8**:*36*, 38
Huygens, Christaan **2**:19, 20
hydraulics **3**:26, 30
hydrochloric acid **8**:*12*, 13, 16–18
hydrogen **3**:8; **8**:15; **9**:10
hydrogen sulfide **8**:*12*, 13
hydrophone **7**:28
hydroxides **8**:*12*, 13, 19
hypodermic needle **8**:*21*

I

IBM Research Laboratory **6**:25
image intensifier **6**:*16*
Improved TOS satellites **6**:18
inches **1**:6–7, 13, 29
index glass **1**:22
India **7**:*37*
inductance **4**:8
inertia **3**:6
Infrared Astronomical Satellite (IRAS) **6**:*18*, 19
infrared radiation **6**:*11*, 14–15
detectors **5**:36; **6**:8, 15–16, *18*, 19
heat **6**:8, 10
production of **6**:16
satellites **6**:18–19
spectrum **6**:*17*, 18; **8**:*23*
submillimeter **6**:18
infrasound
animals **7**:*38*, 39
earthquakes **7**:32, 35, *36*, 36–37
spiders **7**:*38*
insects **7**:26; **9**:7, *9*, 24, 25, 29
insulators **4**:7
interference **5**:10, 12, *12*
interferometer **6**:*38*, 39
International Bureau of Weights and Measures **3**:6
International Date Line **1**:36; **2**:*38*
International Ultraviolet Explorer **6**:13
invertebrates **9**:25, 28
iodine **9**:11, 12, 13
ionic bonds **3**:*32*, 32–33, *33*
ions
bonds **3**:33, *33*
charged **4**:13; **8**:15, **8**:26–27
electrolysis **4**:*18*
electrons **8**:26–27
helium **3**:36
metals **4**:13; **8**:*12*
iron hydroxide **8**:19, *19*
iron oxide **8**:19, *19*
isobars **1**:37
isotopes **2**:34, **3**:38–39, *39*
infrared lasers **6**:16
radioactivity **2**:34; **3**:39
radiocarbon dating **2**:36–37

J

Jansky, Karl **6**:36
Japanese Subaru Telescope **6**:19
JNWP weather forecasting **8**:39
joules **4**:32
Jupiter **3**:15; **5**:22; **9**:38, *38–39*

K

keyboards **7**:*23*
kidney stones **7**:25
kilogram per cubic meter **3**:8
kilograms **3**:6
kilometers **1**:*28*, *24*, 6, 13
kiloparsecs **1**:19
kilopascals **3**:8
kilovolts **4**:8
kilowatt-hours **4**:33, *34*
kilowatts **4**:9, 32
kingdoms **9**:*32*
Kirchhoff, Gustav **8**:*22*
Kitt Peak National Observatory **5**:32

L

laboratory **9**:*11*
Landsat **1**:*37*; **6**:*18*; **1**:*36*
Landsteiner, Karl **9**:30, 31, *31*
Langevin, Paul **7**:29
Langley, Samuel Pierpont **6**:15–16
lanthanides **9**:*12*
Large Electron Positron Collider **3**:37–38, *38*
lasers **1**:14, *15*; **6**:16; **7**:19, *19*
latitude **1**:22, 34–35, 36; **2**:17, *28*
lawrencium **9**:*12*
lead **8**:21
lead sulfide **6**:16
leadsman **7**:28
leap seconds **2**:39
leap years **2**:8
leaves **9**:19, *19*
LEDs **6**:16
legumes **9**:20, *20*
length **1**:6–7, 13
lenses **5**:14–15, 27, *27*
cameras **5**:34–35
concave **5**:*14*, 14–15, 26
condenser **5**:38, *38*
convex **5**:*14*, 14–15, *15*, 22, 25
diverging **5**:*14*, 15, 26
electron **6**:24
erector **5**:27, *27*
eyepiece **5**:16, *17*, *18*, 20, 23, 24, 25–26
fish-eye **5**:36
magnetic **6**:21
objective **5**:16, *17*, *18*, 19, 23–24, 25–26
telephoto **5**:36, *37*
ultrawide-angle **5**:36
Van Leeuwenhoek **5**:*21*
wide-angle **5**:35–36
see also microscopes; optical instruments; telescopes
leopard **1**:*10*; **9**:*24*
Libby, Willard **2**:*36*, *37*
library classification **9**:8–9
light
beam **5**:8, 10, 11
colors **5**:8–9, *8–9*
diffraction **3**:35, *35*; **5**:10, 11, *11*
focus **5**:14–15
haze **6**:14
infrared **6**:18–19
interference **5**:10, 12, *12*
lenses **5**:14–15
microscopes **5**:*17*, 18–19, *20*
radiation **5**:7–8; **6**:6
rays **5**:*14*, 15, 28
refraction **5**:8, 9, 10, 14
speed **5**:8; **6**:9
ultraviolet **6**:*6*, 7, 13; **8**:31
waves **5**:6–9; **8**:24
white **5**:*6*, 7, *9*, 10, 12, 25
light radar (LIDAR) **6**:16
light years **1**:13, 19
lightning **4**:*13*
limestone **2**:*34*; **9**:17
line spectra **8**:*22*, 24
line workers **4**:*27*
linear accelerators **3**:37
Linnaeus, Carolus **9**:18, 23, *23*
Linné, Karl von **9**:*23*
lion **1**:*10*; **9**:26, *27*
Lippershey, Hans **5**:22, 26
liquid hydrogen **3**:37
liquids **3**:9; **8**:24
lithium **8**:*22*
location **1**:38, *38*, 39
longitude **1**:34–35, 36; **2**:26–28, 38–39
loudness **7**:6, 7, *7*
loudspeakers **7**:20–23, *21*, *22*

Lowell Observatory **5**:*31*
lutetium **9**:*12*
Lycophytes **9**:22

M

Mach, Ernst **7**:9
Mach number **7**:9
McLeod, Herbert **3**:21
McMath-Pierce Solar Telescope **5**:32
magic lantern **5**:38, 39
maglev train **4**:*36*, 37
magma **9**:17
magnetic detectors **4**:37–38
magnetic north **1**:39
magnetic resonance imaging **8**:26
magnetic tape **7**:16–18, *17*
magnetism
  magnetic field **4**:14, 15, 16–17, 18, *18*, 36–39
  poles **4**:36, 37
  repulsion **4**:17, *17*
magnetometer **4**:*37*, 37–38, *39*
magnetron **6**:*32*, 32–33
magnification
  binoculars **5**:27
  focal length **5**:23–24
  microscopes **5**:14, *20*; **6**:22, 24
  telescopes **5**:23–24
magnifying glass **5**:14, 15, *15*
Magnoliophyta **9**:20, 23
mammals **9**:*25*, *26*, *27*, 28
  marsupials **9**:25
  monotremes **9**:25
  placental **9**:25–27
manometer **3**:20, *20*, *21*
maps **1**:*14*, 36–37
  compass **1**:*38*, *39*
  coordinates **1**:35
  distance **1**:32
  location **1**:*38*, 39
  projections **1**:37
  scale **1**:32
  shading **1**:*35*, 37
  symbols **1**:*32–33*
marble **9**:17
marching bands **7**:*6*
Marrison, Warren A. **2**:24
Mars **3**:25; **9**:38, *38–39*
marsupials **9**:25
Maskelyne, Nevil **2**:28
mass **3**:6–7, 8, 11–12
  gravity **3**:11
  inertia **3**:6
  measures **3**:16
  see also balances
mass spectrometers **8**:20, *21*
mass spectroscopy **8**:26, *27*
mathematics **8**:32
Mauna Kea, Hawaii **5**:*30*, 32; **6**:19
mean **8**:33
measuring tapes **1**:8
median **8**:33
megahertz **6**:33; **7**:26
megaparsec **1**:19
megapascals **3**:8
megatonnes **3**:6
megavolts **4**:8
megawatts **4**:9, 32
Mendeleev, Dmitri **9**:10, 11, 12, *13*
mendelevium **9**:*13*
Mercalli, Giuseppe **7**:*37*
Mercator, Gerardus **1**:37
Mercury **3**:25; **6**:10; **9**:36, 38, *38–39*
mercury **8**:*22*
  barometer **3**:23, *24*
  fluorescent tube **6**:12, *12*
Merops ornatus **9**:*28*
mesons **3**:36, 37
metal detectors **4**:37–38, *38*
metals **8**:*12*, 13
  alkali **8**:19
  carbonates **4**:13; **8**:*12*, 14, 19
  chlorides **8**:13
  hydroxides **8**:*12*, 13, 19
  identifying **8**:20–21
  inspection **6**:27
  ions **4**:13; **8**:*12*
meteorology *see* weather forecasting
meteors **9**:38
meter bridge **4**:26, *26*
meters **1**:6–7, 13
methyl group **8**:25
Mexico City **7**:*36*
microanalysis **8**:13
microbalance **3**:*17*, 17–18
microfarads **4**:8
micrograms **3**:6
micrograph, scanning electron **6**:*20*, *23*, *24*
micrometer gauge **1**:*11*, 12, *12*
micro-meters **1**:12, *13*
microorganisms **5**:16, 21
microphones **7**:10
  amplifier **7**:16
  carbon **7**:14, *14*
  condenser **7**:12, *12*
  crystal **7**:12
  diaphragm **7**:10, 11, 12, 14
  dynamic **7**:11
  hydrophone **7**:28
  moving-coil **7**:11, *11*
  ribbon **7**:11
  telephones **7**:10, 11, 14
  transducer **7**:10
microscopes
  bacteria **5**:16, 21
  biology **5**:16, 21
  condenser **5**:*17*, 19
  focus **5**:20
  geology **5**:21
  lenses **3**:*24*; **5**:16, *17*, *18*, 18–20
  light **5**:*17*, 18–19, *20*
  magnification **5**:14, *20*; **6**:22, 24
  mechanical stage **5**:20
  mirror **5**:*17*, *18*
  specimen **6**:21
  turret **5**:*17*, 19
  Van Leeuwenhoek **5**:*21*
microscopes, types
  atomic force **6**:24
  binocular **5**:*16*, 20
  compound **5**:*16*, 16–20, *17*, *18*
  electron **3**:33; **6**:*20*, 20–25
  scanning electron **6**:*22*, 22–23
  scanning probe **6**:23–24
  scanning tunneling **6**:23, 25
  simple **5**:16, *21*
  transmission electron **6**:*21*, 21–22, 23, *25*
microwave ovens **6**:*9*, 33
microwaves **6**:33
  cellphone network **6**:34–35, *34–35*
  echoes **6**:33–34
  radiation **6**:19
  satellites **6**:33; **7**:15
  water **6**:33
  wavelengths **6**:6, 8–9, *9*, 10, *11*, 19, 32–33
miles **1**:6, 13, 14, 16, *28*
Milky Way **6**:18, 36, 38; **9**:36
milliammeter **4**:29
milliamperes **4**:7, 18
millibars **3**:9
milligrams **3**:6
millimeters **1**:6, 10, 13; **1**:24
millivolts **4**:8
milliwatt **4**:9
milometer **1**:16–17
minerals **8**:*11*; **9**:14, 16
  aerial photographs **6**:15
  artificial earthquakes **7**:*39*
  classification **9**:*15*, *16*, 16–17
  combinations **9**:17
  extracted for use **9**:*15*, 17
  hardness **9**:16–17
minibeasts **9**:*6–7*, 7–8
  see also bugs
minutes, angles **1**:20
minutes, time **2**:9
mirror
  microscopes **5**:*17*, *18*
  periscopes **5**:27
  telescopes **5**:28–29, 30, 32; **6**:19
mites **9**:8
mode **8**:33
Mohs, Friedrich **9**:16
Mohs scale of hardness **9**:16–17
molar solution **8**:9
mole **8**:9
molecular biology **3**:35
molecules **3**:32, **8**:25, 28
mollusks **9**:25, 28
Monera kingdom **9**:32, *32*
monkeys **9**:26, 31
Mono Lake **5**:*37*
monococcus **9**:33, *33*
monocotyledon **9**:21, 23
monotremes **9**:25
months **2**:8–9, 10
Moog, Robert **7**:23
Moon **1**:18; **4**:*33*
  calendars **2**:11
  gravity **3**:15, 18
  longitude **2**:27–28
  lunar tables **2**:28
  months **2**:8–9
  orbit **3**:12
  telescopes **5**:22, *23*
mosquito **6**:*24*
moths **9**:8, *9*
motion, laws of **3**:*13*
motorbike **1**:*17*
Mount Wilson Observatory **9**:*39*
multimeter **4**:29, 30, *31*
music **5**:12; **7**:9, 23

N

nanometer **5**:7; **6**:26
Napoleon **4**:*19*
NASA **1**:*36*; **6**:13, 18, 19
National Bureau of Standards **2**:23
National Institute of Standards and Technology **2**:23
Naval Research Laboratory **2**:24
navigation **1**:22; **7**:26
nebula **6**:14, *15*; **9**:*34*, 36
Neptune **1**:18; **9**:38, *38–39*
neutrinos **3**:36
neutrons **2**:34; **3**:36, *36*, 37, 38–39
New Brunswick **3**:13
New Mexico **6**:*37*, *38*
Newlands, John **9**:11
newton **3**:6, 8, 18
Newton, Sir Isaac **3**:6, 11, *13*; **5**:10, *29*
  see also reflecting telescopes
newtonmeter **3**:18
Nicole, Adolphi **2**:29
Nile River floods **3**:13–14
nilometer **3**:14
Nilson, Lars **9**:12
nitrogen atoms **3**:36, *37*, 38
NMR spectroscopy **8**:25–26, *26*
noise pollution **7**:*9*
Norway **1**:*15*
Nova Scotia **3**:13
nuclear fission **3**:39
nuclear magnetic resonance see NMR
nuclear reactor **3**:39, *39*
nucleus
  atoms **2**:34, **3**:36, *36*; **6**:12; **8**:23
  protons **9**:13
  radiation **8**:25–26
numerical weather prediction models **8**:38–39
nuts **9**:20, *20*

O

observatories **5**:24, *30*, *31*, 32; **6**:18, 19; **9**:*39*
odometer **1**:16
Ohm, Georg **4**:6, 8, *31*
ohmmeter **4**:29, 29–30
ohms **4**:8, 9, 20
oil deposits **1**:*15*
open-pit strip mine **8**:*11*
opera glasses **5**:26
optical instruments **5**:14, 15, *15*
  binoculars **5**:*26*, 26–27
  cameras **5**:34–37
  periscopes **5**:27
  projectors **5**:38–39, *38–39*
  see also microscopes; telescopes
optical spectroscope **8**:*22*
orbit **2**:6; **3**:12, 15; **9**:36
orders **9**:25, 26
organic compounds **8**:24, 25, 26
Orion **9**:*34*
oscillator **7**:26
oscilloscope **4**:21–23, *23*, 25
osmosis **3**:*30*, 30–31, *31*
osmotic pressure **3**:*27*
ounces **3**:6
overhead projectors **5**:39, *39*
oxides **8**:19
oxygen **3**:*37*, 38; **8**:*18*; **9**:11
ozone layer **6**:10, *11*

P

Pantheon **1**:*24*
parallax **1**:*18*, 18–19
parallelograms **1**:25, *25*
parasites **9**:33
parenthood testing **8**:*30*, 30–31
Paria Wilderness Area **2**:*35*
parsec **1**:18, 19
particle accelerators **3**:37
Pascal, Blaise **8**:*35*
pascals **3**:8, *24*; **8**:*35*
pedometer **1**:*16*
pendulum **2**:*18*, 18–19; **5**:*24*
Penning, Frans **3**:21
pepo **9**:20, *20*
perambulator **1**:16
Periodic Table **8**:*12*; **9**:10, 11–13, *12–13*

Set Index

periscopes **5**:27, *27*
permeability **3**:30–31
phonograph **4**:*35*
phosphor **4**:22; **6**:12
phosphorescent screen **6**:21–22
photoconductors **6**:16
photography
aerial **6**:15, *17*
earthquakes **7**:*35, 36*
infrared **6**:*14*, 14–15, 18
time-lapse **5**:*37*
wildlife **5**:36
x-rays **6**:28
see also cameras
photosynthesis **2**:36–37; **9**:18
phylum **9**:24, 25, *25*, 28, *32*
pi **1**:26, 29
picofarads **4**:8
picoseconds **2**:31
pictogram **8**:*33, 34*, 35
pie chart **8**:*33*, 35
Pierce, George Washington **2**:25
piezoelectricity **2**:25; **7**:12, 24
Pinwheel Galaxy **1**:*18*
pipette **8**:*16*, 17, *17*
Pirani, Marcello **3**:21
pit viper **6**:8
pitch **7**:6, *7*, 9, 30–31
Pitot, Henri de **3**:22
Pitot tube **3**:22
placental mammals **9**:25–26
planetariums **5**:38–39
planets **1**:18; **9**:*34*, 34–35, 38
plankton **5**:*19*
plants **9**:*18, 32*
classification **9**:18–23
naming system **9**:23
plate movement **7**:32, *32, 33*
platinum **4**:30; **6**:16; **8**:20–21
Pluto **9**:38, *38–39*
Polaris **1**:19
pollution **5**:33; **7**:*9*
polymerase chain reaction **8**:30
pome **9**:20, *20*
positron emission tomography (PET) **6**:30, *31*
potassium **2**:36; **8**:21, *22*; **9**:11
potassium-40 **2**:34, 36
potassium permanganate **8**:18
potential difference **4**:7–8, 9, 20
potential divider **4**:28, *28*
potentiometer **4**:20–21, *22*
pounds **3**:6
precipitate **8**:10, *10, 12*, 19, *19*
precipitation reactions **8**:*12*, 13, 18
pregnancy **7**:*24*, 25, *25, 26*
pressure **3**:8–9
depth **3**:9
gases **3**:20–25
liquid **3**:9, 26–31
osmotic **3**:*31*
see also atmospheric pressure
pressure gauges
Bourdon **3**:21, *23*
McLeod **3**:21–22
Penning **3**:21
Pirani **3**:21
primates **9**:26, *26*
prime meridian **1**:35, 36; **2**:26, 27, 39, *39*
prism, light **5**:10, 11, *11*; **8**:22
prism, solid **1**:29
probability **8**:*34*, 35, *35*
projections, maps **1**:32–34
azimuthal **1**:34, *34*
conical **1**:34, *34*
cylindrical **1**:34, *34, 37*
Mercator **1**:34, *37*
projectors **5**:38–39, *38–39*
protons **2**:34; **3**:36, *36*, 37, 38; **9**:13
protractor **1**:*20*, 21, *21*
Psilophyta **9**:22–23
Puerto Rico **6**:*39*
pumps
centrifugal **3**:29
force **3**:26, *26*, 29
impeller **3**:29
lift **3**:25, *28–29*, 28–30
piston **3**:26
rotary **3**:29
see also heart
purification of mixtures **8**:*10*
pyramid **1**:*29, 31*

Q

quarks **3**:36, *36*, 37
quartz **6**:13; **7**:12; **9**:16, 17
quartz fiber **4**:*11*, 12
quasars **6**:38

R

radar **6**:*32*, 33–34
radiation
spectroscopy **8**:20, 24–27
vacuum **5**:7; **6**:9; **7**:6
see also gamma rays; infrared; light; ultraviolet; x-rays
radio astronomy **6**:10, 36, *37*, 38
Radio Corporation of America **6**:25
radio receivers **8**:*32*
radio telescopes **6**:36–39, *37*
radio transmitters **2**:25; **6**:32
radio waves **5**:7; **6**:6; **8**:26
Sun **6**:38
wavelengths **6**:8–9, *9*, 10, *11*, 36
radioactivity
gamma rays **6**:6–7, 27
isotopes **2**:34; **3**:39
radiocarbon dating **2**:36–37, *37*
radiographer **6**:*27*
radiology **6**:28, *28*, 29, *31*
radiometric dating **2**:34
radiotherapy treatment **6**:26, 29
radium **9**:*12*
rainbow **5**:*6, 8, 9*, 9, 10; **6**:15
rat-bite fever **9**:33
recording studios **7**:12
rectangles **1**:*25*, 25
red giants **9**:36
redox titrations **8**:18
redwood tree **3**:*27*
reflecting telescopes
Cassegrain telescope **5**:*29*, 29–30
Coudé telescope **5**:30
Gregorian telescope **5**:29
Hale telescope **5**:29
Herschelian telescope **5**:28–29
Hubble space telescope **5**:33, *33*; **6**:13; **9**:39
infrared astronomy **6**:*18*, 19
Keck telescopes **5**:32
Newtonian telescope **5**:28–29, *29*
Schmidt-Cassegrain telescope **5**:30
Schmidt telescope **5**:30
solar telescope **5**:32, *32*
WIYN **5**:*28*
see also observatories
refraction *see* light
relative atomic mass **9**:10, 11
relative molecular mass **8**:9
reptiles **9**:25, *25*
resistance **4**:26–31
ammeter **4**:15, 20, *20*
conductors **4**:26
electrical **4**:7, *27*; **6**:16
ohm **4**:8, 9, 20
restriction enzymes **8**:29
revolver **8**:*9*
Rhynchocephalia **9**:25
Richter, Charles **7**:*37*
rickets **6**:7
rock band **7**:*23*
rock salt **3**:*33*
rockets **2**:30; **3**:*14*, 25; **6**:18
rocks
ages of **2**:36
igneous **2**:34, 36; **9**:17, *17*
metamorphic **9**:17, *17*
sedimentary **9**:17, *17*
slices **5**:21
strata **9**:*14*
roentgen rays **6**:26
Rohrer, Heinrich **6**:25
Romans **2**:14, 16–17
Röntgen, Wilhelm Konrad **6**:26, *26*, 28
rubidium **8**:23
ruby **9**:14
Rudbeckia hirta **9**:*21*, 23
rulers **1**:8
Ruska, Ernst **6**:24–25
Russell, Henry **9**:35
Russian royal family **8**:31
Rutherford, Ernest **3**:36

S

safety glasses **8**:*20*
samara, double **9**:20, *20*
sampling **7**:17–18
San Andreas Fault **7**:34
sandglass **2**:12
sandstone **2**:*34, 35*; **9**:17
sapphire **9**:14
satellites **1**:*36*; **6**:10, *17*
infrared radiation **6**:15, 18–19
Landsat **1**:*36, 37*; **6**:18
location **1**:38, 39
microwaves **6**:33; **7**:15
Rosat **6**:13
SPOT-1 **6**:*17*, 18
UV **6**:13
Saturn **3**:15; **6**:*15, 36*; **9**:38, *38–39*
scale **1**:32; **4**:*16*, 17, *17*
scale bar **1**:32, 41
scandium **9**:12
schist **2**:*34*
Schmidt, Bernhard **5**:30
scouring rush **9**:22
seabed **7**:*31*
seahorses **6**:*7*; **9**:26
seasons **2**:6, *6, 7*
seconds, angles **1**:20
seconds, time **2**:6, 9, 23
security alarms **6**:15
seismographs **7**:32, *34*, 34–35, *35, 38*
selenium **9**:10, 11
semiconductors **6**:16
Sequoia National Park **3**:*27*
sextant **1**:22, *22*; **2**:26, *28*
shadow clock **2**:15–16
shale **2**:*34*
shot-putting **1**:*8*
shunt **4**:20
shutter, cameras **5**:35–36
Siemens AG **6**:25
silicon **6**:16; **9**:12
silver **4**:6
silver birch **9**:*22*
silver chloride **8**:19
silver nitrate **8**:19
Sirius **1**:19
skin cancer **6**:7, 10
skulls **9**:27, 28
slide **5**:17, *18*
slide projector **5**:*38*, 38–39
snakes **6**:8; **9**:25, *25*
snow crystals **3**:*34*
soap bubbles **5**:11, *13*
Socorro **6**:*37, 38*
sodium **8**:20–21, *22*, 26–27; **9**:11
sodium chloride **3**:33, *33*
soil **9**:18
solar cells **4**:*14, 33*
solar day **2**:8, *8*
solar mass **3**:7
Solar System **1**:18; **3**:15; **9**:38
solids, volume **1**:28–31
solstices **2**:6, *7*, 8
solution **3**:30, **8**:9, *10*, 12, *16*
solvent **3**:30; **8**:*10*
sonar
dunking **7**:31
fishing **7**:*28–29*, 30–31
passive **7**:31
submarine detection **2**:25; **7**:29–30
sound
detection **7**:10–15
intensity **7**:7
loudness **7**:6
pitch **7**:6, *7*, 9
speed **7**:8–9
synthetic **7**:23
see also ultrasonic sound
sound recording **7**:16–19
sound reproduction **7**:20–23
sound signals **7**:15, *15*
sound waves **7**:6–7, *7*
diffraction **5**:12
earthquakes **7**:32
frequency **7**:6–9, *8*
interference **5**:12
traveling medium **7**:6; 8–9
sounding **7**:28
space program **5**:34
Space Shuttle **3**:14–15; **5**:33; **9**:*39*
spacecraft **3**:15
see also rockets
Spain **2**:15
spark plugs **1**:11
speakers see loudspeakers
species **9**:18, 26
specific gravity **3**:8, 41
specimen **5**:*17*, 18, 19, 21; **6**:21
SPECT scan **6**:*29*
spectroscopy **8**:6–7, *20*
astronomy **8**:7, 22–23
infrared **8**:*23*, 24–25
NMR **8**:25–26, *26*

  optical 8:22
  radiation 8:20
  ultraviolet 8:24–25, 25
spectrum 5:7, 8
  infrared 6:18; 8:23
  prisms 5:10, 11
  rainbow 6:15
  Sun 8:22
  ultraviolet 6:13
speed 2:32
  air 3:22
  cameras 2:31
  car 2:32
  light 5:8; 6:9
  limits 2:32
  sound 7:8–9
speedometer 1:16, 17; 2:32, 32
spheres 1:26, 26, 29
Spherophyta 9:22
sphygmomanometer 3:30, 31
spiders 7:38; 9:29
spirilla 9:33, 33
spirochete 9:33, 33
Spirogyra 5:20
sports 1:8; 2:30–31
SPOT-1 6:17, 18
spot heights 1:37
spotlights 4:28
spreadsheets 8:37
Squamata 9:25
squares 1:24, 25, 25
stadium 1:19
Stanford Linear Collider 3:37–38
staphylococcus 9:33, 33
stars 9:34, 34–35, 36
  brightness-temperature relationship 9:35–36
  constellations 2:11; 9:35
  mass 3:7
  radiation 6:15–16; 9:38
  x-rays 9:38
starting blocks 2:30–31
static electricity 4:10, 10, 23
statistics 8:8–9, 33, 33, 34, 35
stereo systems 7:21
stereoscopic vision 5:26, 27
sterilization 6:27
stopclock 2:28
stopwatch 2:28–29, 31
storm damage 8:36, 38
strain gauges 2:25
streptococcus 9:33, 33
strontium 8:14, 21; 9:11
strontium salt 8:20
Su Sung 2:14–15
subatomic particles 3:36, 37
submarines 2:25; 3:9, 7:29–30, 30
Sun 1:18; 2:6; 5:32
  calendars 2:10
  constellations 2:11; 9:34–35
  eclipse 5:37
  evolving 9:36
  gamma rays 6:10, 11
  heat rays 6:8, 10, 14
  infrared radiation 6:14
  light 5:7; 6:7
  orbiting bodies 9:38, 39
  photosynthesis 9:18
  radio waves 6:38
  rainbow 5:9
  spectrum 8:22
  spots on surface 5:22
  ultraviolet radiation 6:7, 10
  x-rays 6:10, 11
sundials 2:12, 15–16, 17
supercomputer 8:38
supergiants 9:36
supernovas 6:38; 9:36
surveyor's chain 1:14
suspension 8:10
swimming 2:31
swimming pool 1:7
switch 4:6
synchronization 7:19
synchrotron 3:37
synthesizers 7:23
syphilis 9:33

T

tables 8:33
tachometer 2:32–33, 33
tape measures 1:8
tape recorders 7:16–18, 17
Taraxacum officinale 9:21
teeth 1:10; 9:24, 26–27, 27
telephones 7:9
  cables 7:15
  cellphones 6:34–35, 34–35
  microphones 7:10, 11, 14
  microwave satellite links 7:15
telescopes 5:22–24, 28
  binocular 5:26
  galaxies 9:36
  Galilean 5:24, 25, 25–26
  lenses 3:24, 5:14
  mirror 5:28–29, 30, 32; 6:19
  mountings 5:24, 25
  radio telescopes 6:36–39, 37, 39
  refracting 5:22, 22–27, 23
  terrestrial 5:23, 27
  ultraviolet 6:13
  see also reflecting telescopes
Television and Infra Red Observation Satellite (TIROS) 6:18
tellurium 9:11–12, 13
tennis court 1:6–7, 7
test meter 4:29, 31
theodolite 1:14, 15
thermal conductivity 3:21
thermocouple 6:16
thermograms 6:14–15, 19
thermometers 4:30
Thompson, J. J. 3:36
thorium 9:10
Thoth 2:13
throwing events 1:8
thunderstorm 7:8–9
tides 3:12, 12–14
tiger 9:24, 26
time
  Earth 2:34–37
  elapsed 2:26–29
  prime meridian 2:39, 39
  units 2:6, 9
  universal 2:39
  see also clocks; sundials
time base 4:22
time zones 2:38–39
timepieces 2:12
timing of events 2:28–30
Tiros Operational System (TOS) 6:18
titration 8:17, 18
tomography 7:25
  x-rays 6:30, 30, 31
Tompion, Thomas 2:20
tonnes 3:6
tons 3:6
tornadoes 8:38, 39
torr 3:24
Torriano, Juanelo 2:21
Torricelli, Evangelista 3:23–24, 24
tracers, radioactivity 6:30, 31
transducers 7:10, 21, 24
transformers 4:35; 7:22
triangles 1:25, 25
trigonometry 1:19, 21
Tsvet, Mikhail 8:15
tuatara 9:25
tumors 6:14–15, 26; 7:25
tungsten 4:7; 6:23
turbines 3:14; 4:9
tweeters 7:21

U

UK Admiralty 2:26, 27–28, 30
ultrasonic sound
  animals 7:26, 27
  crystal vibrations 2:25
  navigation 7:26
  pitch 7:30–31
  submarine detection 7:29–30, 30
  underwater echoes 7:28–31
  uses 7:24, 24–27, 25, 26
  welds 7:26, 27
ultrasound scanner 7:24, 24–25, 25, 26
ultraviolet radiation 6:7, 11, 12
  production of 6:10, 12–13
  spectrum 6:13
underwater life 5:34
units
  analysis 8:9
  astronomical 1:13, 14, 18, 19
  customary 1:6; 3:6, 8, 9
  density 3:8
  electricity 4:6–9
  length 1:6–7, 13
  mass 3:6–7
  metric 1:6
  pressure 3:8–9
  time 2:6, 29
uranium 3:38; 6:13; 9:12
uranium-235 3:39, 39
uranium isotope 2:34; 3:38, 39
Uranus 6:15; 9:38, 38–39
U.S. Bureau of Standards 3:6

V

vacuum, radiation 5:7; 6:9; 7:6
vacuum tube 6:32
Van Leeuwenhoek, Anton 5:21
Venus 2:10; 3:25; 5:22; 9:36, 38, 38–39
Vernier, Pierre 1:10
Very Large Array 6:37, 38
vibrio 9:33, 33
volcanoes 2:34
Volta, Alessandro 4:6, 8, 23
voltage 4:6, 9
  amplifier 4:12
  battery 4:21, 22
  measuring 4:20–25
  potential difference 4:7–8, 20
voltaic pile 4:23
voltameter 4:12, 13
voltmeters 4:20, 20, 21, 23
volts 4:8, 9
volume 1:28, 29, 29, 30, 30, 31
Von Laue, Max 3:35

W

Warren, Henry 2:21
watches 2:21, 24–25
water
  acidity testing 8:18
  density 3:8
  depth 7:28
  microwaves 6:33
  molecules 3:32
  oxygen 8:18
  pressure 3:26–31
  volume 1:28
  see also underwater echoes
water vapor 6:18
Watt, James 4:9, 32
wattmeter 4:32, 32–33
watts 4:9, 32
wavelength
  colors 5:6, 7
  gamma rays 6:6–7
  light 5:6–9; 8:24
  microwaves 6:6, 8–9, 9, 10, 11, 19, 32–33
  radiation 6:6–9, 19
  radio waves 6:8–9, 9, 10, 11, 36
  sound waves 7:6–7, 7
  spectrum 5:7, 8
  ultraviolet radiation 6:7
  x-rays 6:6–7, 24, 26
waves
  amplitude 5:7, 12
  crests 7:6–7, 7
  frequency 5:7
  troughs 7:6–7, 7
  see also wavelength
waywiser 1:17
weather forecasting 1:23, 23; 8:9, 36, 36, 37, 38–39
weather maps 1:37
weathering 9:17, 17, 18
weighbridge 3:19
weight 3:6–7, 16
weightlifter 3:7
welds 7:26, 27
wells 3:28
whales 7:26, 31
wheel-cutting engine 2:21
white dwarfs 9:36
Whitworth, Joseph 2:21
Wiener, Alexander 9:31
Wight, Isle of 6:17
wind erosion 2:35; 9:14, 17, 17
Winkler, Clemens 9:12
wire gauges 1:9, 11
woofers 7:21
worms 9:28
Wyatt, John 3:19
Wyck, Henry de 2:18

X

x-ray crystallography 3:35
x-rays 6:7, 10, 11
  dangers 4:12; 7:24
  detection 6:28
  electrometer 4:11, 12
  production 6:26–27
  stars 9:38
  tomography 6:30, 30, 31
  uses 6:26–27, 27, 28, 29, 29
  wavelength 6:6–7, 24, 26
x-ray spectrum analyzer 6:24
x-ray tube 6:26, 27, 27, 30
Xi Cygni star 9:34

Y

yards 1:6, 16
yaws 9:33
Yeager, Chuck 7:9
year 2:6, 9
Yerkes Observatory 5:24; 9:39

Z

zodiac 2:11
zoology 9:26–28
Zworykin, Vladimir 6:25

Further reading/websites and picture credits

Further Reading

Atoms and Molecules by Philip Roxbee-Cox; E D C Publications, 1992.

Electricity and Magnetism (Smart Science) by Robert Sneddon; Heinemann, 1999.

Electronic Communication (Hello Out There) by Chris Oxlade; Franklin Watts, 1998.

Energy (Science Concepts) by Alvin Silverstein et al.; Twenty First Century, 1998.

A Handbook to the Universe: Explanations of Matter, Energy, Space, and Time for Beginning Scientific Thinkers by Richard Paul; Chicago Review Press, 1993.

Heat (How Things Work Series) by Andrew Dunn; Thomson Learning, 1992.

How Things Work: The Physics of Everyday Life by Louis A. Bloomfield; John Wiley & Sons, 2001.

Introduction to Light: The Physics of Light, Vision and Color by Gary Waldman; Dover Publications, 2002.

Light and Optics (Science) by Allan B. Cobb; Rosen Publishing Group, 2000.

Electricity and Magnetism (Fascinating Science Projects) by Bobbi Searle; Copper Beech Books, 2002.

Basic Physics: A Self-Teaching Guide by Karl F. Kuhm; John Wiley & Sons, 1996.

Eyewitness Visual Dictionaries: Physics by Jack Challoner; DK Publishing, 1995.

Makers of Science by Michael Allaby and Derek Gjertsen; Oxford University Press, 2002.

Physics Matters by John O.E. Clark et al.; Grolier Educational, 2001.

Science and Technology by Lisa Watts; E D C/Usborne, 1995.

Sound (Make It Work! Science) by Wendy Baker, John Barnes (Illustrator); Two-Can Publishing, 2000.

Websites

Astronomy questions and answers — http://www.allexperts.com/getExpert.asp?Category=1360

Blood classification — http://sln.fi.edu/biosci/blood/types.html

Chemical elements — http://www.chemicalelements.com

Using and handling data — http://www.mathsisfun.com/data.html

Diffraction grating — http://hyperphysics.phy-astr.gsu.edu/hbase/phyopt/grating.html

How things work — http://rabi.phys.virginia.edu/HTW/

Pressure — http://ldaps.ivv.nasa.gov/Physics/pressure.html

About rainbows — http://unidata.ucar.edu/staff/blynds/rnbow.html

Story of the Richter Scale — http://www.dkonline.com/science/private/earthquest/contents/hall2.html

The rock cycle — http://www.schoolchem.com/rk1.htm

Views of the solar system — http://www.solarviews.com/eng/homepage.htm

The physics of sound — http://www.glenbrook.k12.il.us/gbssci/phys/Class/sound/u11l2c.html

A definition of mass spectrometry — http://www.sciex.com/products/aboutmass.htm

Walk through time. The evolution of time measurement — http://physics.nist.gov/GenInt/Time/time.html

How does ultrasound work? — http://www.imaginiscorp.com/ultrasound/index.asp?mode=1

X-ray astronomy — http://www.xray.mpe.mpg.de/

Picture Credits

Abbreviation: SPL Science Photo Library

6 Odile Noel/Redferns; **9t** E. de Malglaive/The Flight Collection; **9b** Library of Congress/SPL; **10** AGE Fotostock/ImageState; **13** Michael Ochs Archives/Redferns; **16** Lew Long/Corbis Stock Market; **20** Granville Harris/ImageState; **23** Fin Costello/Redferns; **24** Ruth Jenkinson/MIDIRS/SPL; **26** Dept. of Clinical Radiology, Salisbury District Hospital/SPL; **26–27** Merlin Tuttle/SPL; **30, 30** inset Photo Press/The Defence Picture Library; **31t, 31b** W. Haxby, Lamont-Doherty Earth Observatory/SPL; **35t** David Parker/SPL; **35b** Tom McHugh/SPL; **36** David Leah/SPL; **38** Dr. John Brackenbury/SPL.